D0394003

THE WILD AND THE WICKED

Benjamin Hale

THE WILD AND THE WICKED

On Nature and Human Nature

The MIT Press
Cambridge, Massachusetts
London, England

This book was set in Heron Sans and Century Schoolbook Pro by Toppan Best-set Premedia Limited. Printed on recycled paper and bound in the United States of America.

Library of Congress Cataloging-in-Publication Data

Names: Hale, Benjamin, 1972- author.
Title: The wild and the wicked : on nature and human nature / Benjamin Hale.
Description: Cambridge, MA : MIT Press, [2016] | Includes bibliographical references and index.
Identifiers: LCCN 2016018207 | ISBN 9780262035408 (hardcover : alk. paper)
Subjects: LCSH: Environmentalism–Philosophy. | Philosophy of nature. | Nature–Effect of human beings on.
Classification: LCC GE195 .H34 2016 | DDC 304.2–dc23 LC record available at https://lccn.loc.gov/2016018207

10 9 8 7 6 5 4 3 2 1

CONTENTS

ACKNOWLEDGMENTS

This book is very much an experiment, an attempt to integrate the complicated technical analysis of contemporary philosophy into a manuscript that is accessible to an educated, but perhaps not philosophically informed, readership. The problem is that to merge the abstract, disinterested, and careful reflection of philosophy with the snarly, politicized, and convoluted practical considerations of our environmental quagmire has meant making more than a few sacrifices in technical specificity, taking quite a few risks with style and voice, and entertaining scenarios that otherwise don't jive with well-accepted political commitments. It has also meant suffering the slings and arrows of a referee process that is more geared toward academic precision than readability. Frankly, it has been more rigorous and demanding than any I've experienced with other publications.

There are many, no doubt, who will take great umbrage with what I've written here. Indeed, readers and referees of this manuscript have been of two minds: either they love it or they hate it. Some referees have taken me to the mats over the looseness of my argument. Others have criticized the glibness of my prose. Still others have bludgeoned me because they sense that the views I discuss here are politically dangerous, giving too much credence to the other side. It's easy to pick nits. For this, then, I should acknowledge my referees and my critics, but not in a "thank you" kind of way.

Though I have my detractors, I also have my supporters, and it is this unbelievable collection of friends and family that has kept me going through what has been one of the most trying publication efforts of my life. All of the people I have encountered and crossed paths with over the past several decades of environmental activism have influenced me immensely, and I owe each of them a shout-out for their support and effort in trying to make the world a better place. I mention them briefly in this book: Schwibby, Mookie, Oyster, Badger, Turtle, Dirt, Mad Dog, and others. Mad props to you all for your work to improve the planet.

The many who read and provided comments on early and later versions of this manuscript deserve attention as well: Roger Fortuna, Henry Pickford, Chad Kautzer, Simon Sparks, Allen Thompson, Ken Shockley, Elissa Guralnick, Brian Calvert, and actually quite a few others have been invaluable critics and supportive friends.

I should also thank some of my extremely close friends and professional colleagues. My dear friend and mentor, Andrew Light, provided for me immeasurable support and encouragement throughout this project. I've learned a great deal from him about environmental ethics, academic politics, and climate policy. I also owe him a great debt, as I think we all do, for his tireless work helping to craft a viable international climate agreement. My Bizarro doppelgänger, Alastair Norcross, through his interminable ribbing, has galvanized for me my conviction that consequentialism cannot do the work that many believe it capable of. I suppose more than anyone, he's my imaginary interlocutor here. After all these years, I have no hope of convincing him, but I'm pretty sure he's never hoped to convince me either. I should also thank another close friend and colleague, Roger Pielke Jr., who has helped me see much more clearly how foggy and confusing climate policy and climate politics can be. He's been encouraging and helpful along this whole journey. We've shared many beers together, and I've always found him to be a constructive and encouraging critic.

As I've slowly banged away at drafts of this manuscript, the team at the MIT Press has been incredibly supportive and understanding. Beth Clevenger is wise beyond her years. Her professionalism and enthusiasm made the submission and production process much less painful than it could have been.

But no acknowledgments section would be complete without attending to those who have attended to me in the kindest, most important ways. My mother, Amy Hale, for teaching me to give a shit and supporting me through the more uncertain times. I'll never forget sheepishly knocking on her bedroom door at 3:30 a.m.—[tap-tap-tap], "Mom?"—to confess my dark and painful secret: I was going to major in philosophy. My mother's partner, Jack Lynn, for his lifelong commitment to conservation and for his willingness to become my son's "Gaia Father," assuming the burden of teaching him to live responsibly on this planet. My little sister, Lauren Hale, an accomplished academic in her own right, offered for me a role model of academic success. Though I'm four years her elder, I've always felt that I'm playing a game of catch-up. My father also played a role in the development of this book, mostly through my imagined conversations with him about why he should vote for such-and-such a candidate.

If it is appropriate to thank the deceased, I suspect I owe no greater debt than that which I owe to my uncle, Jesse Mantel. Though he died of a glioblastoma when I was 18, he was alive long enough to instill in me a lasting interest in the deep questions. Not only was he eminently well-read, but he introduced me to thinkers I would later come to relish. His struggle with cancer, perhaps, was my first introduction to the brutality and unfairness of nature.

Above all, my wife, Becky McNeil, has helped me take this book from seedling to fruit. Not only has she read multiple versions of this manuscript and helped me cautiously navigate the publication gauntlet, she's also provided emotional and moral support all along the way. She is without question my best friend and main pillar. For at least the past ten years, we've been blessed by our son, Jasper, who has reintroduced the world to us and helped to take the edge off our cynicism. I am reminded, as I watch him grow, why we should care about our planet in the first place. It is my fervent hope that as we think more carefully about our environmental obligations we can find a way to give him some peace and stability when he has children of his own.

INTRODUCTION

Apart from the time I substituted a cup of salt for a cup of sugar, one of my first cooking misadventures occurred in the early 1990s, on a Thanksgiving trip home from college. Filled with an arsenal of ideas and a mind for social change, I was home to proselytize, eager to persuade my family that a vegetarian diet was not only the right diet but the tastiest diet as well. It was my view then that each of us is obligated to do our part to undermine the negative impacts of factory farms. Not only are such farms cruel to animals, I thought, but they are also an extremely inefficient way of providing food. I was of the mind that each American bore the burden to change his or her behavior to help put factory farms out of business. And so I took it upon myself to demonstrate that a good, healthy holiday meal needn't be propped up by honey-baked hams and richly stuffed turkeys.

The avoidable disaster, I believe, was attributable as much to my undeveloped culinary ability as to my ideological exuberance. My fatal mistake? Serving the tofu raw.

Of those around the table, my mother gave the most visceral feedback. Where I so eagerly hoped that the meal would prove persuasive, I remember watching her closely for any indication of resistance or rejection. She cooperated fabulously. "Looks delicious!" she said. Taking her fork, she first poked and then scooped up a sizable triangle of tofu, clamping her lips around it with confidence. Moments later her eyes shifted from side to side, her cheeks grew concave, and she vigorously though valiantly forced a smile and a nod. Casually, she brought a napkin to her lips as if to wipe them. It was clear to me, at least, that what had moments ago entered her mouth as something delicious was now being smuggled out in a tidy little paper package.

So much for tofu.

Since then I've sharpened my game considerably. Indeed, as an environmental ethicist at one of the premier environmental research universities in the United States, I've focused my research on arguments that aim to justify environmentalism—why we should preserve, defend, or protect nature in the face of so many other challenges like poverty, hunger, injustice, disease, and so on. Over time I've noticed something pretty remarkable: Many people think exactly as I used to. They think that in order to argue that environmentalism is right, they must also show that nature is good, that *being green* is good. Their thought, then, is that to be green, one should embrace all things natural. One should buy granola and drive a hybrid and drink tea and cook tofu. In effect, one should start hugging trees. Unfortunately, this view belies a deep-seated intuition that nature is not as precious as some claim, which in turn begets a *big* problem. Antienvironmentalists are left with a gargantuan opening: What if it turns out that nature isn't the precious Eden that some environmentalists claim? What if, as many suspect, nature has a dark side?

Consider this: In the wee hours of a warm September night, not long after my wife and I moved to Colorado from New York City, I lay in a deep sleep in my bedroom at the mouth of Boulder Canyon. If you're not familiar with the geography of Colorado, Boulder is a bustling university town at the foot of the Rocky Mountains. It is flanked on one side by hundreds of miles of prairie and on the other side by a mountain range that shoots up out of the prairie just four blocks from where my house then stood. Long before daybreak, I was startled awake by a violent, visceral screech outside my open window. The sounds sent chills through my skin. I heard chesty growling, bones crunching, bark tearing from a tree, and a gurgling shriek so prolonged that I half wondered whether I had woken from a nightmare or whether my nightmare had just begun. Within a few moments the sounds diminished to nearly nothing. I heard only muddled grunts, whimpers, and snorts, peppered with that memorable chesty growl. As if by instinct, rather than hiding under the covers to secure my safety, I leapt from the bed, approached the window, and stuck my face to the screen. What I saw, or what I think I saw, in vague outline, was a bobcat tearing the flesh from a now limp, now twitching, small animal; a squirrel probably, but possibly a house cat. I recoiled then, in mild horror.

Was this event bad? Was this good? Was it right? Or wrong? Was it *evil*? It was disgusting, to be sure. Terrifying, in a way. But was it right or wrong?

Nature is a cruel mistress. She will send her minions to gobble up your pets, sweep away your belongings in the middle of the night, assault your neighbor with a terminal disease, and sucker punch your family members with a dose of their mortality so brutal and cold that you may wonder whether life is worth living. She sends us hurricanes, tornadoes, earthquakes, and tsunamis; cancer, plagues, bird flu, Ebola, malaria, typhoid, meningitis, and even rogue, killer asteroids to pull the dark sheets of death over life just as quickly and fecklessly as she will, with the other hand, breathe life into our newborns and our gardens. It is therefore appalling, to some, that anyone would ever deign to be an environmentalist, that anyone would ever deign to love nature.

I suspect that a great number of people have been turned off from being green exactly for the reason that my mother was turned off from tofu and vegetarianism. Though greening one's lifestyle may be appealing in principle, making dramatic changes can be uncomfortable, off-putting, and even unpleasant. A very intuitive way of trying to combat this discomfort is to batten down the hatches, to strengthen one's commitment to the value of nature. Many nascent environmentalists think that if they just *insist* that nature is something to cherish, that it is something we should adore, then everything else will fall into line. Everyone will suddenly do the right and the green thing. This would be fine, except that once we get to the airport to find that our flight has been delayed thanks to a snowstorm, or we arrive at the ballpark to find that the game has been rained out, the harsh realities of nature suddenly seem extraordinarily inconvenient. So this is where it gets very, very sticky.

It is my view that no matter how much one reveres nature, no matter how much one loves sunsets and beautiful vistas, no matter how poetically one makes the case for preserving this green globe of ours, it is an ongoing struggle to remember these feelings when making day-to-day decisions. The reason for this? Tofu is terrible. Tents are cold. Ticks and termites are a nightmare. Most of us beneficiaries of modern industrial civilization say "no thanks" to these things. Give me a movie, a beer, and a bowl of popcorn any day.

My aim in this book is to argue that environmentalists needn't be so concerned to defend the value of nature; that we needn't grow weary when opponents mock us for being idealistic sissies; indeed, that we needn't necessarily even adopt the attitude of tree-hugging nature lovers. The problem for environmentalists isn't one of isolating and defending the value in nature. The problem is that many non-environmentalists don't see the *need* to isolate the value in nature. Many non-environmentalists think that their actions can be justified by appeal to other ends—money, jobs, welfare, lives, etc. Put a little differently, the problem for the environmentalist isn't a problem of having the right values; it's a problem of having the right theory of justification. Here I offer such a theory.

I argue this position by taking as my starting point exactly the opposite presumption—the idea not that nature is grand and wonderful and awesome, as many of us believe it to be, but instead that nature is nasty and horrible and cruel. My ending point, my conclusion, will be that even if nature is cruel, we still need to be environmentally conscientious in our actions and policies. It's not that I necessarily believe nature to be so cruel, it's just that adopting this strategy will help me illustrate my fundamental point: that we don't need to love nature to be green. We needn't necessarily extol nature's wonders, I will argue, because what really ought to be driving our environmentalism is our humanity, not nature's value.

NATURE'S VALUE, THE CHOICES WE MAKE, AND THE LIVES WE LIVE

There should be no illusions: this argument about nature's value, about nature's preciousness, remains at the center of it all.[1] Anybody who has ever gotten a solicitation letter from the World Wildlife Foundation, or the Nature Conservancy, or the Sierra Club, or any of the many other related conservation organizations, is familiar with the awed and reverent tones often used to describe the natural environment. Anybody who has ever hunched in the corner of a bookstore, sipping coffee underneath a silvery Ansel Adams print, will recognize the same sentiment. They know, deep down, the amount of ink that the luminaries of environmentalism—authors, politicians, business owners, park rangers, outdoor enthusiasts, and filmmakers—have spilled seeking to demonstrate how superb and fantastic nature is, how beautiful it is, presumably to persuade others that there is magic there. Thoreau's *Walden*, Leopold's *A Sand County Almanac*, as well as works by Ralph Waldo Emerson, John Muir, Thomas Berry, Annie Dillard, Diane Ackerman, James Lovelock, and John McPhee, among many others, all carry strains of this romanticism. Even Joyce Kilmer's famous poem begins "I think that I shall never see a poem as lovely as a tree"—a schmaltzy serenade if there ever was one. A good portion of the literature characterizes nature as the most precious gift in the world, as if polar bears and baby seals, as if penguins and prairie dogs, as if snail darters and spotted owls, were the very kin of Bambi.

Oh sure, many of these authors are also realistic about the dark side of nature. Many, in fact, struggle with the tensions between the beautiful, the wild, and the horrific. But the fact remains that many, if not most, of them emphasize, and in some cases romanticize, the beauty of nature, presumably in hope that you too will be carried along with them in their appreciation for the environment. The irony here is that it is precisely this romantic picture of nature that turns many otherwise good and well-meaning people *away* from environmentalism. Its cloying sappiness has never been successful at masking the significant perils of nature. How is one to be persuaded by a peripatetic glass-half-fuller when one's mother dies of cancer or one's child dies of malaria?

Lest you think that I am starting from a bad place by picking fights with environmentalists, it may be important to identify the romantic picture's most pernicious incarnation: not its appearance in the canonical environmental literature, but its prominence in politicized caricatures of environmentalists. The examples are legion. Radio and television personalities Rush Limbaugh and

Glenn Beck routinely depict "environmentalist wackos" as those who think of humanity as a "disease" and would rather return us to the Dark Ages than to permit technology to move us forward.[2] *Wall Street Journal* columnist Bret Stephens has suggested that the climate crisis is a liberal phantom that distracts from much more pressing issues like terrorism. Senator Ted Cruz has claimed that "global warming alarmists are the equivalent of flat-earthers."[3] Karl Rove, in response to the 2015 Paris Accord, proposed that idealistic environmentalists are insensitive to the plight of the developing world: "What we are now saying to emerging economies [is], 'Keep your people poor, keep them in poverty because you cannot use cheap fuels, namely natural gas, coal and other fossil fuels to power your economy.' And it's ridiculous." And even the slightly less polarizing (!) *Washington Post* columnist Charles Krauthammer once implied that environmentalists so value nature that they are willing to sacrifice the health and well-being of humans over the needs of caribou and seals.[4] And this is just the beginning. Many politicians and pundits before and after these folks have forged campaigns by spinning the suggestion that the environmental community fetishizes wildlife and nature to the negligence of economic considerations.

Environmentalists, of course, are well aware of this tension. We've been struggling to find a foothold in the political discourse since *Silent Spring*. Our current alternative—making the case that each of us has a personal interest in ensuring nature's survival—isn't much different from its early instantiations. Climate change offers a particularly simple example of the problem that comes with affixing our long-term self-interest to our current practices. As it has been for decades, the emphasis is on the value of nature—specifically, its value for our survival. Strategically, it seems to be working. It's abundantly obvious that the threat of climate change has motivated millions to think more conscientiously about the way they live. Should be a good thing, right? But ethically and politically, these are much hazier gains.

Ethically, the environmental community's preoccupation with climate change has diverted most of the discussion from the concern for nature that initially inspired environmentalists. As environmentalists take their eye off the ball and fret over alternatives to energy consumption, animal and plant species continue to disappear. Less evident is that since climate change is primarily a problem with emissions, it appears to require only a technical solution—and in the current case we've turned to focusing attention mostly on carbon reduction. Meanwhile, this technical emphasis invites extraordinary global environmental interventions like geoengineering—radical proposals to steer the earth's climate toward stability—and subverts some of the founding impulses of the environmental movement.

Politically, it's also a problem. Boiling down a complex ethical problem to a simpler Chicken Little–style campaign opens the door for simple political rejoinders. Where Paul Crutzen and Eugene Störmer coined the term *Anthropocene* to describe the current interval of the Holocene epoch during which human activities have left their mark on the planet's ecosystems, many have responded in ways that do not please environmentalists. Well-positioned conservation biologists like Peter Kareiva and Michelle Marvier have developed a "New Conservation Science," which employs the tactic of triage to prioritize species that benefit humans. Emma Marris, in her fantastic book *Rambunctious Garden*, suggests that we should rethink our relationship to nature, abandon the idea of the wild as pure and untrammeled, and instead reconceive of ourselves as nature's gardeners. The new "Ecomodernist Manifesto"—spearheaded by Michael Shellenberger and Ted Nordhaus, but endorsed also by enthusiastic geoengineer David Keith, my colleague Roger Pielke Jr., philosopher Mark Sagoff, and many important others—suggests abandoning the traditional environmentalist insistence that humans harmonize with nature and in its stead adopting a more "pragmatic," technology-embracing, future-forward environmentalism.

Such redirection rightly worries environmentalists, as these proposals threaten to upend the few gains that have been made through the climate change debate. These proposals are all at once a simple call to reorient the objectives of environmentalism and a challenge to commitments that many environmental stalwarts hold dear. The instinctual step for many is to return to old and tired tropes—to try even harder to tie global environmental collapse to the precious value of nature. Witness the weepy pictures of lone polar bears desperately clutching remnant ice floes. Far from a satisfactory response to the critics of environmentalism, however, this strategy is the same hackneyed appeal that environmentalists have been making for decades—an appeal to the value and fragility of nature—in an effort to bring the rest of the world around to doing the right thing.

Fortunately, there is an alternative. There's no need to throw the baby out with the bathwater. We can accept many of the commitments that have motivated environmentalists for decades without insisting that we prioritize human interests, that we become nature's gardeners, or that we embrace technological solutions to our problems. To go this way, however, it's important for environmentalists to stop emphasizing the value of nature and to focus instead on our ethical obligations to justify our actions. It is important for the environmental community to acknowledge that nature isn't necessarily all that it's cracked up to be; for us to acknowledge that, at times, nature can be pretty brutal. Nature is a mixed bag. Many times nature can be good … for us, for others; many times nature can be just plain awful. This may sound like heresy, but if you feel this way, I hope that you will hear me out. Like many people, I consider myself an environmentalist, and I have taken many actions, and written many papers, in defense of nature.

Do I love nature? In a way, I suppose. I've even hugged a tree or two in my time. But I've certainly been miserable while camping, and I most definitely don't love mosquitoes, feeling water in my boots, or tasting sand in my stew. And I don't like snow, or cold, or bad weather. I've cursed cats when they've scratched me and I've even smashed a few spiders with my shoe. I hate cancer, and I have a vintage bottle of wine saved up for the day on which the disease is eradicated from the earth. But environmentalism is not about love. And I don't think it's fundamentally about self-interest either. I think it's about reason and rationality. It's about being human.

We should be environmentalists because it is *right*, because we have the capacity to be better than nature. What I mean by this is that we human beings have the unique power to be moral, to justify our actions, to evaluate the reasons that we have for acting, and *then* to take action. We're not like lions or rats or pigs.[5] We're human—a fact that carries with it special burdens. Namely, human beings can guide their actions by reason; they can justify their actions. According to the same logic that has us chastising people who "act like animals," we ought to recognize that it is our responsibility *not* to act like animals. In order to do this, we need to have reasons behind us, *good* reasons. A man who wanders into the woods and kills a dog just to see it die, we might rightly condemn as loony, beastly, or immoral. But we might feel differently if that same man kills a dog that is attacking him. His reasons, we might say, are justified. Our unique burden as human beings is that we can act for reasons, good or bad, and thus we fail to live up to our potential if we act as brutally as nature. These reasons I speak of? They inform everything we do.

Consider the variety of choices we face. Consider, in particular, the way in which we reason through these choices. What should we do? Should we buy a hybrid or a gas guzzler? On one hand, buying a hybrid seems like the right thing to do. No one wants to be a part of the climate change problem. On the other hand, for many people the cost of a hybrid is prohibitive. It would be much cheaper and easier to buy the $3,000 used car advertised on Craigslist. Is there anything wrong with this sort of calculation?

Consider also our choices at the grocery store. There, on the shelf, are several cartons of eggs. One is cheap, at $1.49. Another is twice as expensive. FREE-RANGE ORGANIC, a label screams from the more expensive carton, soy-ink practically bleeding through the cardboard. It's a small extra cost, by itself, but it's significant when paired with the pricey organic milk right beside it, as well as the many other things we must purchase on any given shopping trip. Why should we choose one over the other?

Most of the time we get very little guidance in such decisions. Either we are left to our own devices, basing our choices on our private repository of knowledge and desires, or we are steered by others, basing our choices on paths laid out by self-anointed experts: our parents, our friends, the billboard down the street, the hotshot how-to-green-your-lifestyle manual of the year, or the trendsetting superstar. We hold fast to the idea that our decisions are ours to make, our actions are ours to take. But this is the perplexing and inconvenient conundrum of environmentalism: our decisions and actions impact an enormous and global spectrum of others, human and nonhuman alike. If we take time to consider the implications of our decisions before we even get to the grocery store, before we force-feed ourselves disastrous meals of unseasoned and uncooked tofu, we will be in a much better position to say that we have lived an ethical life.

Many environmentalists operate under the mistaken presumption that their arguments for the value of nature, if made well and strongly enough, will through sheer force of reason persuade other nonenvironmentalists to become environmentalists. Much to their chagrin, this generally doesn't pan out. The reason for this, I believe, is that the justificatory train has already left the station. In other words, the problem for environmentalists isn't that people don't value nature; it's that many people don't see the relationship between the value of nature and the justifiability of the actions they take. Many people seem to think that an action is justified so long as it fulfills a need, a want, it is lawful, or they have the resources to make it happen. "It's a free country," after all. This default position stems from a particular set of ideas about human freedom, not about value. Therefore, if environmentalists are sincere in their efforts to change how people think and act, what must be advanced instead is not another theory propounding the value of nature, but a theory of justification that is not rooted in value—a theory that explains why we humans must scrutinize and analyze our actions to be sure that we are not destroying our world. That's what I offer here.

WHAT GREEN MEANS

Over the past few years, thanks in part to Al Gore and the International Panel on Climate Change, as well as innovative green entrepreneurs like Yvon Chouinard (founder of Patagonia), Jeffrey Hollender (founder of Seventh Generation), and Jeff Lebesch and Kim Jordan (founders of New Belgium Brewing Company)—who have made it easier to live comfortably while being green—we've seen an explosion of interest in environmentalism. Seems like every business now has an environmental angle. Some of us in the environmental community no doubt cheer this turn of events—it's what we've been striving for. But there's also a sense among those of us who remember being derisively sneered at as "tree huggers" that all this interest in being green is as flimsy and fleeting as an interest in hula hoops. The enthusiastic chants of "Drill, baby, drill!" at the 2008 Republican National Convention only reinforce our fears.

In fact, the cracks in the environmental edifice are already showing. Even though much of the talk of the Green New Deal centers on stimulating a green, alternative energy economy, it's not clear that individuals and families, strapped as they often are, will find it within themselves to make lasting changes. Since this is where most people think the rubber hits the environmental road, this is worrisome indeed. So what are we to do? Are we to ramp up the rhetoric? To pursue the line of reasoning that has dominated environmentalism for decades? Or is there some other way?

As I've said, I think there's another way.

In a particularly candid offhand remark, then-presidential candidate Barack Obama perhaps put it best: "we can't solve global warming because I f—ing changed light bulbs in my house. It's because of something collective."[6]

I'm inclined to agree with Obama, though not exactly for the reasons that he may have in mind. I agree that being green isn't characterized by a particular set of private actions on the part of any individual. Many of our most vexing environmental problems are problems for all of us; that result from the actions of all of us. Since this is the case, they require a collective response. What this collective response should look like, however, is not as straightforward as the question about which kinds of behaviors are earth-loving and which are not. Rather, it means that we have to pursue collectively valuable ends, and we have to coordinate our actions at the social level.

The best way to understand this is through our choices, via the reasons that we tacitly endorse each time we choose a course of action. Each one of us makes environmentally relevant choices all the time. It's insane to think that any single principle will be broad enough and at the same time nuanced enough to suit any one of us regarding all of our separate and personal choices. Instead, we must strive for a comprehensive strategy that impacts the decisions of all people, so that the particular concerns of private individuals are not weighed down with considerations about what is tastiest, easiest, and most convenient. This means that environmental problems must be answered with a renewed call to think, to justify our actions.

We humans are in a unique position to be better and less destructive than nature. We can do so, I argue, if we take our job as reasoning, essentially philosophical, beings seriously. This book presents an argument based not on squishy emotional commitments but rather on the moral demands of being a vulnerable human on a harsh planet.

Far from what you might expect, my stance is not to demonize nature. It's nothing so simple. Nature is neither good nor bad. It is a force indifferent to morality, a force that pushes along regardless of human interruption. Often this force can be beautiful, stunning, and wonderful, just as many nature writers suggest. But just as often it can be merciless and cruel. And yet despite nature's indifference to the plight of humanity, humanity cannot be indifferent to the plight of nature. Because we are able to choose how to act, and have the ability to apply reason to our behavior, we should celebrate our humanity, relish our differences from nature, all the while recognizing that our humanity comes with a heavy burden: to ensure that our actions are justified.

An orientation toward justification should inform the decisions that we all make daily: what to eat, what to buy, what to drive, where to live, how to live, where to invest, whom to vote for. This book lays out the case that we must, each, be environmentalists, even if we remain agnostic on, or in some cases antagonistic to, the value of nature.

ABOUT THE BOOK

The Wild and the Wicked is a chronicle of sorts, traveling backward and forward through events in time—moments of horror, calamity, chaos, and crisis. It is an exploration of our duty to respect nature, in spite of nature's manifest disregard of, and frequent cruelty toward, us. It is also an argument for an alternative approach to environmentalism. As I make my case, I'll offer different and new answers to common questions about choices that we all make—choices about family, diet, business, politics, and even national security. I arrive at my final destination—that yes, we are morally obligated to be green—by way of a distinction common in moral philosophy: the distinction between the good and the right.

The distinction itself is simple, but perhaps best understood in its negative form. When something awful in nature happens, this is *bad*. When people do bad things to one another, this is *wrong*. In personal, political, and business discourse, the two are frequently confused, so much so that they color our conception of what we are obligated to do. Yet the distinction between the good and the right remains one of the most powerful in moral philosophy, allowing us to talk of responsibility, blame, guilt, and authority. Commingled with the elements of ethical theory, the distinction can serve as a platform from which to rethink entirely a human relationship to nature. The good-right distinction prompts us to examine questions about evil, utopia, justice, forces of nature, and respect for others, culminating in the claim that one has an obligation to be green.

The underlying question at the heart of this book is therefore considerably broader than an innocuous interest in nature. It is the subject matter of the discipline in which I have made my home. It is this: How should we live? This question has been asked in various forms by great thinkers since the beginning of recorded time: Socrates, Plato, Aristotle, Augustine, Aquinas, Hobbes, Bentham, Mill, Rousseau, and on up the intellectual's ladder; each has asked this seemingly simple question. How should we live? It is the fundamental question of ethics, interpreted by some philosophers as a question specifically about what we should do, and interpreted by other philosophers as a more personal question about what kind of person we should be.

Until recently, it was generally thought to pertain expressly to human beings: How should we treat one another? But not anymore. Now we have a problem. Our problem is that nature is bearing down on us—or, more accurately, we are bearing down on it. The environmental problem of which I speak is not limited to global climate change, though you wouldn't know it from news reports. It's much more pervasive than that. Our population is growing. Our standard of living is rising. Our neighbors are getting richer. We are getting richer—even in spite of recent economic turmoil. And as all of this happens, nature makes her limits and her brutality far more palpably known. As our population grows, we demand more food. As our standard of living rises, we use more raw materials. As we and our neighbors grow richer, we demand more space. As we live our lives in the 21st century, we encroach on habitat, we pollute rivers and streams, and we consume our nonrenewable resources to the point of exhaustion. In short, we bump up against nature; and as we do, nature pushes back. With every year—as we get richer, as we consume more space—the costs to us of nature's damage grow steeper, and the costs to nature from us grow irreversible.

It's an unsustainable slamdance. Like a crushing throng of fans at a punk rock concert, as our numbers and demands grow, we squeeze our population and hopes and dreams into the retaining walls of nature. As we do so, nature squeezes back, tightening around us like a chain-link fence pressed to our cheeks. So is it now for us: we build our homes on floodplains, in the foothills of mountain ranges, on fault lines, at the mouths of volcanoes, in the middle of swamps, in the paths of hurricanes, and so on; and as we do, we face threats that before we had kept at bay. Mountain lions hang out in our backyards (or, at least, in my backyard), hurricanes threaten our coastal regions, earthquakes challenge our cities, and so on and so on and so on. Time and again nature repeats its destructive dance, tearing asunder the ramparts that we have built up. Just turn on the news and you will undoubtedly witness startling images of some poor soul bailing out his home while nature bears down around him. In this, our advanced industrial civilization, more people have more to lose; and nature just does her thing. Simultaneously, the more people do their thing, the more nature loses.

Long ago, perhaps, when the world was so vast that conquistadors and explorers would bravely risk their lives to venture into unknown territories, the idea of a limit on the earth's capacity was unfathomable. It would have seemed crazy, back then, to suggest that one should conserve wood because there are a limited number of trees in the forest, or that one should conserve fish because there are a limited number of fish in the sea. Not so now. Nature's limits are in view. Resources are shipped from the farthest reaches of the earth to support the lives of those who live in sprawling suburbs, miles from city centers, miles from one another; ocean ecosystems and fisheries collapse under unsustainable harvesting practices; shadow economies that traffic in addiction, exploitation, and prostitution—in short, misery—spring up to support these industries; and we burn oil, oil, oil, as though we've never quite gotten the picture that *nonrenewable* will someday mean "the well is dry."

So what are we to do? Are we to sit idly while this happens? Are we to bury our heads in the sand? Or are we, like so many others before us, to revere nature for her beauty, tranquility, and magnificence, in hope that others will catch on and start valuing the world around them?

Not a chance. Nature is a bitch. But that's not even close to the whole story.

What to Expect

Whatever the origins of the view that nature is the bee's knees, this book takes aim directly at the heart of this view. The book therefore touches on many instances, stories, and tales that emphasize the nastiness of nature and the ugliness of humankind. My intent is that these tales will shed light on real-world practical questions on the true meaning of *green*. Like many good chronicles, the book comes in several acts, each oriented to give you a clearer sense of how to handle problems that you face every day. Act I (chapters 1–4) clearly lays out the environmental landscape, taking special note of the frame of the debate and the central motivating concerns for many environmentalists. Act II (chapters 5–7) demonstrates how this frame informs our answers to questions about how to be green and explains how a minor philosophical adjustment can better position us to address these problems. Act III (chapters 8–10) introduces a new paradigm and political orientation for the green movement: the Viridian Commonwealth. In the end I suggest approaching environmental problems by leaning into two central philosophical notions: (1) that the idea of the good is manifestly confusing when it comes to our responsibilities to one another, and (2) that the best way to understand our obligation to the environment is by understanding the full range of questions tied up in a robust consideration of the right. Don't worry if you don't understand this now. There's much more to say.

I proceed by addressing calamities large and small, at the individual level and at the global level, natural and human-caused. I talk about natural disasters, animal attacks, and deadly epidemics. I discuss the freakish creatures of mythology and legend, like the chupacabra, the sand squink, and the fur-bearing trout. I cover several well-known natural calamities, including the Boxing Day Tsunami of 2004, the extinction of the dinosaurs, and the case of a young boy who was mauled to death by a bear; as well as instances of human-caused nightmares, ranging from the bombing of Hiroshima to the 9/11 attacks to climate change. In the end, it is not the calamities that interest me but rather the nature of the relationship between human beings and those calamities. To underscore the philosophical dimensions of this discussion, I draw several bizarre analogies. At one point I even discuss the relationship between our uses of language and scrambled eggs. You'll have to read the first part of the book to see that these analogies are not as irrelevant as they may seem.

Some of the suggestions here will be familiar to many of you, though there will also be many that you may never have heard before. Some have made headlines and gotten a great deal of attention, whereas others have slid under the radar. I'm not an historian or an anthropologist, so I hope you don't hold me to the standards of those fields. I'm trained primarily as a philosopher. As such, I don't trade in facts; I trade in ideas. The purpose of the tales is not to introduce you to an obscure event in the history of the world. My hope is that these familiar and unfamiliar stories start you thinking about nature and environmentalism in a way that you haven't thought about them before.

I've written this book to be as entertaining as it is informative. With the exception of those who have, in fact, borne the brunt of nature's wrath, everybody loves a disaster story. The reason for that, I suspect, is that disasters challenge us to think about our own mortality. So I think there's also an existential query in this book as well. If I'm correct, the analysis provided here will help you make clearer sense of almost all choices and circumstances in your lives, including those circumstances that involve decisions about how to treat nature. Having said this, I should reiterate that this is not a how-to guide, explaining how to improve your life, but a why-so story. It will provide you with the resources you need to defend yourself against those who think that nature is just ducky as well as from those who think that environmentalism is wrong.

So then, this book really is targeted at you. It is about why you should be as green as the grass—why you should be a responsible environmentalist, regardless of your feelings about nature. It is a book about why you can swat at mosquitoes and crush spiders with your shoe and curse the humidity when it clings oppressively to your skin. From another angle, the concern of this book is the following: As parents, citizens, business owners, teachers, community members, and politicians, must we really be considerate of nature? Yes, I think we must. To establish this point, I'll make the case to you. As you read you should be evaluating my claims and my argument.

Without further ado, I present to you exhibit number one, a trail of wreckage and devastation so heart wrenching that the sympathetic stargazer dare not belabor what the world hath wrought. Ladies and gentlemen of the jury, it is my pleasure to direct your attention to nature—simple, misinterpreted, unadulterated nature. Look at this tangle of thorns.

"The pleasure in this world, it has been said, outweighs the pain; or, at any rate, there is an even balance between the two. If the reader wishes to see shortly whether this statement is true, let him compare the respective feelings of two animals, one of which is engaged in eating the other."

—Professor Schopenhauer, "On the Sufferings of the World"[1]

1

Return to the Paleocene

In which the author (1) introduces the reader to the philosophical significance of extinction, (2) draws a parallel between backward-looking paleontological science and forward-looking climate science, (3) disentangles the political and ethical implications of the climate forecast, (4) establishes the climate problem as centrally of human origin, (5) entices the reader to abandon what she knows, (6) dismantles political considerations regarding the "catastrophe frame," and (7) identifies several problems with understanding our ethical obligations in terms of impending catastrophe.

GONE

It must've come as quite a shock to twelve-year-old Mary Anning when, walking near the Dorset cliffs in England in 1811, she stumbled upon the bones of a dragon.[2] What Anning uncovered, staring hideously up at her from the dirt, were sizable fragments of the spine and skull of an enormous creature many times larger than the fossilized shells she had regularly gathered on that same beach. Anning had already by this time been fossil hunting for several years, helping her family pay the bills by digging up old shells and selling them at market. But those were mere shells. Staring up at her now was something unique and different, something reptilian and crocodile-like—something earth shattering—that soon would pique the curiosity of scientists throughout Europe.

Along with other fossil discoveries from Lyme and Cornwall, each slowly gathered and cataloged since the mid-seventeenth century—the most provocative of which was dubbed the "Scrotum Humanum" for its unchaste resemblance to a giant pair of fossilized human unmentionables—Anning's dragon bones were so extraordinary that they were distributed around the elite scientific circles of Europe, raising new questions about what exactly these things were. Together the bones, gathered not just by Anning but by many others as well, eventually migrated into the hands of the Very Reverend Dr. William Buckland, DD FRS, who in 1824 described and renamed the biggest of the mysterious dragons the *megalosaurus*, and in so doing launched a harried effort to understand the mysterious world that predated, but did not contradict, Noah's flood. Not long thereafter, Buckland's cache of megalosaurus bones was passed along to Sir Richard Owen, a naturalist at the British Museum in London, who renamed the monster a *dinosaur*; and the rest, as they say, is prehistory.[3]

What must have been mind-blowing about the discovery of dinosaurs was the dawning realization that the earth and its inhabitants were quite a bit more fragile than everyone else had previously thought. Imagine Anning's exhilaration, bewilderment, wonder, and terror: There she was, relying primarily on her teachers and parents for her understanding of the world, who themselves were derivatively relying mostly on remnant texts and religious orthodoxy for their understanding of the world, with very little information about those who preceded them, when she must've realized, not long after uncovering these many fossils, that the sand beneath her feet wasn't quite the placid seafront she had once thought.

Once Anning, Buckland, and Owen had uncovered the bones of these dragons, they suddenly had to *explain* those bones. They had to explain what they were, what happened to them, and why the animals from whence they came weren't around anymore. It is one thing to spin yarns of dragon slayers and sea monsters, where those myths are mere figments of one's imagination, where the nature and origin of such creatures remains mysterious and unclear. But to unearth evidence that the ebbs and flows of the universe are such that the megafauna of millennia ago had snouts and scales, teeth and tusks, completely alien from our experiences today, that's a totally different kettle of lungfish.

The prehistory of the planet—literally, the land before time—is a puzzle that we've had to piece together from only the faintest clues, unearthed not over days and weeks, but over decades and centuries. The findings of the past several centuries, passed along through generations of early scientists and paleontologists, have factored into our reconfiguration of the grand timeline. As our friends in the sciences have dug deeper into the soil and the strata, they have gradually offered answers about these first dragons, and they have devised new methods and new techniques that have added observational flesh to the bones of prehistory.

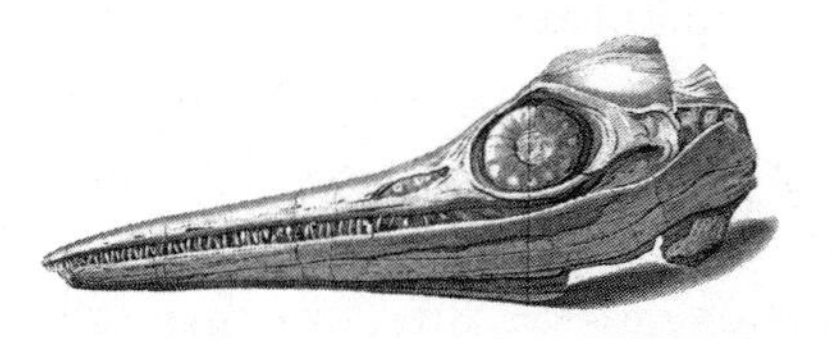

Ichthyosaurus skull. Discovered by Joseph and Mary Anning, 1814.

It may seem a peculiar literary decision to launch a book on environmental responsibility by digging so far back into the history of the world as to lose the relevance of the present, but it is also an undeniable fact about the ascent of humanity, and thereby the emergence of ethics, that humankind is traceable at least as far back as the rise of the mammals, which, in turn, is often linked to the extinction of the dinosaurs.

The big extinction story asserts that approximately 65 million years ago an enormous asteroid smashed into the earth and annihilated a voracious population of great lizards that had theretofore been terrorizing all helpless critters on this otherwise life-abundant planet. The smashing of this asteroid, the story continues, heralded the beginning of a new epoch, the Paleocene, and made possible the proliferation of mammals. The biggest story of all time thus has us believe that once the great predators were gone, the weaker warm-bloods could survive. It's a quaint Darwinian tale, known to most every five-year-old.

But there is another theory about the rise of the mammals, and the subsequent emergence of humanity, that suggests something far more troubling. It goes like this: In the early Paleocene, atmospheric carbon concentrations were roughly 1000 parts per million (ppm). Within the blink of a geological eye—a mere 20,000 years—due to some as yet unexplained terrestrial or extraterrestrial event, carbon concentrations climbed dramatically, forcing the globe to warm by a full six degrees.[4] This transition was abrupt, disruptive, and, in many ways, catastrophic.

It was this perfect sort of climatic upheaval that caused a further mass extinction of 30–50 percent of deep sea biomass, enabling the mammals to grow and flourish in the ensuing tens of millions of years. This period, also known as the Paleocene-Eocene Thermal Maximum, or PETM, was a period of global warming that shares with today not the same atmospheric carbon concentration but the same *rate* of carbon emission into the atmosphere: about 3 percent per year.

Now then, there have been numerous mass extinctions throughout the history of the earth: the Permian, the Triassic, the Jurassic-Cretaceous, not to mention the Late Cambrian, the Late Ordovician, the Late Devonian, and the Paleocene. Since at least the 1980s, many of these have been unreflectively blamed on asteroids—maybe not by scientists, but at least by schoolteachers, Hollywood scriptwriters, and public talking heads—as if the earth were an enormous marble in a fatalistic Pachinko machine. But the PETM is something different. Research strongly indicates that the PETM did not occur from the smashing of an asteroid, but from the gurgling and bubbling of a warming planet due in part to carbon dioxide concentrations in the atmosphere.[5] This connection to the PETM thus opens the door to reevaluate other mass extinctions, and some paleontologists have observed that almost all other mass extinctions exhibit a similar pattern of rapid carbon dioxide emission.[6] The backward-looking extinction stories of yesteryear, in other words, are raising concerns about the forward-looking climate story of today and tomorrow.

The big story of humanity, then, is not so much that the annihilation of the great beasts paved the way for mammalian dominance of this planet, but that a major climatic upheaval, instigated by increasing atmospheric carbon concentrations, brought about such a tremendous turnover in terrestrial life. In a cosmic twist of irony, we are now spewing these selfsame carbon atoms back into the atmosphere, potentially paving the way for us to choke on a new thermal maximum. And that, I think, makes for a good start to a book about the hostility of nature and the humanity of man.

CONNECTING THE DOTS

It would be a mistake, however, to characterize this scientific reconstruction of the early fossil record merely as a herculean project of observational data accumulation. It was also an extraordinary ordeal of conceptual dot-connecting, of inference-making ... of reasoning through discoveries by way of filtering out irrelevant and insignificant noise, by way of abandoning presumption and prejudice, seeking an explanation of the grand trajectory of the universe coherent with the raft of other beliefs that made up the body of knowledge of that day. Reconstructing prehistory from trace fossil evidence involved abandoning and challenging assumptions about the regularity of the universe, introducing notions about upheaval and extinction, but it also involved ascribing physical, chemical, and biological law to these new and sometimes startling discoveries.

It therefore wasn't simply the discovery of dusty old bones that must've blown the minds of those early natural scientists. It was the explosive realization, sewn together by reasonable inferences about the here-and-now, that all of their otherwise solid ideas about the history of the world couldn't hang together as they once thought they could. This revolutionary discovery, probably more than most others, may have given the unsuspecting residents of 19th-century Europe their first whiff of cosmic impermanence.

Acknowledging the impermanence of everyday objects is comparatively easy. It's one of the first concepts we learn as children. Very early in our lives—as infants, say—we apprehend the world first as a series of unrelated events. A ball passes behind a curtain and reemerges on the other side, scaring and delighting us. It's a new ball! Whee! We are bombarded with an enormous range of seemingly unconnected perceptions: red ... here ... now. Over time, however, we learn to organize these perceptions. We infer that the ball passing behind the curtain is the same ball that reemerges, establishing first for us the permanence of objects. This permanence helps us find our footing. It gives us stability. We learn that the ball is identical to the one that came before it, that our room will be the same as it was when we left it moments before, that the people we are talking to now were the same folks with whom we had been talking earlier. As we continue to grow, however, impermanence sneaks back in. We recognize that sometimes our stable expectations can be upended, that balloons pop and cheap toys break and pets die and people pass away. This may not always be a chronological, step-by-step developmental process, but there is an important sense in which the impermanence of objects is tied up in the stability of our tiny little universe.

From appearances it would seem that it is our memory and our recollection of the world moments before that helps us retain the rapid succession of observations and facts as a coherent string of events, punctuated only by our experiences within them. But notice that it's not our memory at all that's doing the organizational work. We ourselves are piecing the whole puzzle together by utilizing our inferential capacity. When a ball goes behind a curtain, it is plausible, though unlikely, that an illusionist, or a child, or a robot, or a delusional chimpanzee with a taste for practical jokes, is catching that ball and rolling a different but identical ball out from the other side of the curtain.

So too with the world. Until we discover evidence that there has been some interruption in the state of affairs, it is natural to assume that there is relatively smooth continuity between the past and the present; it is reasonable to believe that the world that rolls behind a curtain is the same world that rolls out the other side of the curtain. When we do find such evidence, the guardians of knowledge go into a scramble. Wagons get circled, arrows come out of their quivers, and acolytes go on the defensive, much of which has further parallels with the climate shenanigans of today.

• • •

At approximately the same time that Mary Anning and her coterie of fawning intellectual admirers were standing agog at the dragon that would eventually be christened the *icthyosaurus*, naturalist and skeptic Gideon Algernon Mantell, member of the Royal College of Surgeons and fellow of the Royal Society (MRCS FRS), unearthed similarly puzzling and curious old bones, though this time from a physiologically different dragon.

Quite unlike the Very Reverend Buckland DD FRS, Mantell would offer a far more dramatic and substantially different explanation for the origins of his dragon than his colleague to the north, who had theretofore asserted that the bones of Anning's dragon had been washed into a cave of hyenas during a supernatural deluge. Mantell surmised that something else had happened, something terrible, and that the dragon he uncovered had died in an as yet unexplained natural catastrophe. Buckland, who had staked his position on a more religious interpretation of fossil origin, wasn't buying it, and the two, for most of Mantell's career, squabbled over the nature of the find.

When Anning passed along her bones to Buckland, who then passed those bones along to the faculty of Oxford, only to have his findings upended by further discovery and theorizing by heretical ingrates such as Mantell, each fossilist had to radically revise what he otherwise had taken for granted. And this was no simple exercise in sifting through soil or uncovering lost clues as to the origin of the universe, nor was it an exercise in the cultural politics of the scientific establishment. The discovery of dinosaurs was an exercise in argument—in explanatory argument, to be more precise—the oft-overlooked mucilage of the sciences.

This, then, is the first point of the very long argument that I will forge over the course of this book: that our knowledge about the world actually has two sources, not just one. We gain knowledge via both observation and argument. Observation, plainly, is the bread and butter of scientists. It's what they're known for—empiricists, all of them. But reasoning also plays a critical unifying role in the organization and explanation of their factual observations. Thus, contrary to popular belief, it's a misperception that scientists are mere observers of the world. They also rely heavily on argument and inference.

All of this may then help clarify how this book on philosophy is about to take a dramatically different turn from the many other books on science, nature, and politics and that you may have read. Here's what I mean.

This division between fact and argument is rather neatly reflected in the division of academic labor at modern universities. Lab and field scientists tend toward data collection and compilation. They pack up their gear, pull out their instruments, and look for stuff. Theoretical scientists, philosophers, and mathematicians, in contrast, tend toward conceptual argumentation. They read a lot of articles, shut their doors, close their eyes, and think. Many scientists do both—sometimes they observe, sometimes they theorize. As well they should. The two practices don't run in conflict with one another but are, in fact, mutually interdependent.

Think of it this way. Observations without arguments are effectively blind piles of data—accumulated pictures and sensations haphazardly clumped together in a meaningless jumble. Arguments without observations, in contrast, are effectively empty pages of a book, waiting to be filled with details. For a coherent sense of what's going on, both need to work together. The philosopher Immanuel Kant taught us as much over 200 year ago. To quote his famous, but rather cryptic, saying: "Concepts without percepts are empty; percepts without concepts are blind."[7]

Methodologically speaking, we can strengthen our observations by abandoning our conceptual presuppositions, and we can equally well strengthen our understanding of concepts by closing our eyes for a moment. In either case, gaining distance equips us to better understand what's in play. Since I'm an ethicist, most of what I talk about in this book aims at accentuating the strengths and weaknesses of a set of arguments. In this case I assess arguments of a particular variety—normative arguments—which I'll say more about in a moment. First, however, I think we need to take our backward-looking history lesson and turn it to face out across the horizon.

THE FORECAST: UNPLEASANT, WITH A CHANCE OF TERRIBLE

The United Nations Intergovernmental Panel on Climate Change (IPCC), which is the intergovernmental body of scientists responsible for evaluating the risks associated with anthropogenic (or human-caused) climate change, forecasts a range of climatic possibilities, almost all of which are bad. Their fifth assessment report (often referred to as IPCC AR5) anticipates devastation across all trophic zones. Heat-related mortality will increase in Europe. Infectious disease will spread across the Northern Hemisphere. The seas will rise. Coastal regions will incur increasing erosion. Corals will succumb to thermal stress. Wetlands, including salt marshes and mangroves, will be inundated. Millions of people will be flooded out of their homes.[8] Human health will suffer: malnutrition will proliferate, with resulting negative impacts on child growth and development. Many will die from heat waves, floods, storms, fires, and droughts. There will be an increase in ground-level ozone, resulting in cardiorespiratory distress. And that's just a summary of the summary. The full AR5 is hundreds of pages long.

Although catastrophe is not expressly predicted in scientific error bars, it is sometimes trumpeted, and some would say embellished, by political figures and activists who hope to motivate us to change our behavior before it is too late. Messages coming from the United Nations Framework Convention on Climate Change (UNFCCC), which is the primary organizing body for the UN charged with implementing the objectives of the UNFCCC (which is also the name of the international environmental treaty that gives it its name), have become considerably more dramatic, more menacing, more catastrophic. So too from national bodies, like the United Kingdom's Department of Energy and Climate Change and the European Union's Commission on Climate Change. Naturally, the emphasis on climate catastrophe is even hotter coming from activist organizations like Greenpeace or Bill McKibben's 350.org. Nobel Prize winner Al Gore also deserves some credit for ringing the catastrophe bell, but certainly not all. In the eyes of critics he has become the face of the politicization of climate science and stands not as a modern-day Paul Revere but as a symbol of everything that is wrong with environmentalism.

Concern has grown so alarming that some members of the scientific community are joining the political chorus. National Aeronautics and Space Administration (NASA) scientist James Hansen, for instance, has compared climate change to the

Holocaust. Testifying before the utilities board in Iowa, he is quoted as saying, "If we cannot stop the building of more coal-fired power plants, those coal trains will be death trains—no less gruesome than if they were boxcars headed to crematoria, loaded with uncountable irreplaceable species." Hansen has been roundly criticized for this statement, but he nevertheless persists in sounding the alarm.

Joe Romm, a physicist and popular climate blogger with the Center for American Progress, emits a daily fusillade of responses to any developments in climate science or policy. He repeatedly underscores the direness of the climate situation. Whereas Romm responds primarily to policy obstructionists, the many well-respected climate scientists who blog at RealClimate spend hours typing up elaborate explanations of the latest climate science in a heroic attempt to ward off the obfuscation and distortion promulgated by skeptics, contrarians, and denialists. Everything gets knit together. It's all very political. The climate scientists who insist that they are only ever doing pure science—which is to say that they are only reporting the facts, describing the state of the world, and explaining the science—try mightily to distance themselves from the political fray, but they are deeply involved in it.

Don't have time to read the IPCC reports or engage the climate blogosphere? Turn on the television and you will see equally terrifying stuff. My colleague and friend Max Boykoff has conducted an extensive catalog of media reports about climate change by tracking newspaper articles dating back to 1988. He's argued that various journalistic norms of dramatization, novelty, and personalization have contributed to the apparent whirlwind.[9] *New York Times* journalist Andrew Revkin calls this sort of reporting "whiplash journalism." He writes: "the media seem either to overplay a sense of imminent calamity or to ignore the issue altogether because it is not black and white or on a time scale that feels like news. This approach leaves society like a ship at anchor swinging cyclically with the tide and not going anywhere."[10] By almost all accounts, media reports reflect an intense interest in playing up the catastrophe. At the same time, many cable television channels have been airing programs on the devastating forces of nature. Though not directly tied to concerns about climate change, implications about the hostility of nature are no less cautionary.

• • •

There are reasonable, and actually quite terrible, scientific hypotheses about what will happen. There are even important political considerations that can inform our strategizing about what to do. But there is a problem with thinking that the path forward is clear simply because we have a strong scientific basis for the conclusions of climate science. This mistake seems to be made again and again in the environmental community.

Where scientists—and climate scientists in particular—may be assumed by many to be mere reporters of fact, we have already seen that they are also in the business of providing arguments. More important, the explanatory arguments that scientists make are rarely strictly descriptive.

I don't mean this to say that scientific claims are false or problematic. Quite the contrary. They're about the best we can do when it comes to truth. What I mean to say is that they frequently carry a *normative valence*, by which I mean that they appear to imply a prescription as well. They *prescribe* a course of action, much like a physician might prescribe a treatment.[11] In other words, they seem to carry enough information to tell us what we *ought* to do.

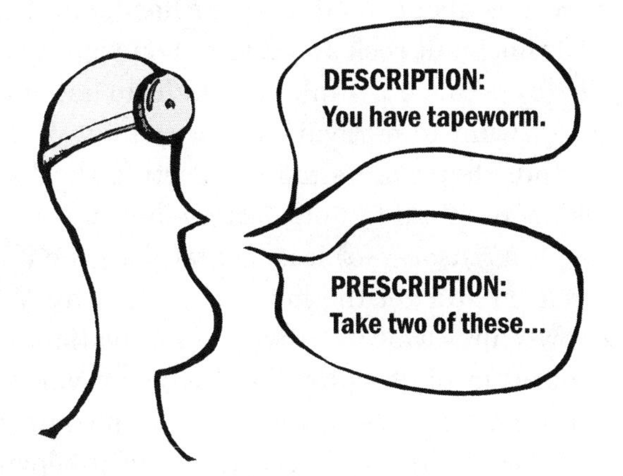

For instance, if I tell you that a rock is about to fall on your head, I am not *merely* reporting the facts to you. I am also imparting a view about that rock and its relation to you: "This will be bad for you. You should move." There is a normative valence in my claim. This goes for almost all factual statements, as well as for full-blown explanatory arguments.

This, then, is my second important point. Some arguments are primarily descriptive, meaning that they describe the world as it is; whereas other arguments are primarily normative, meaning that they offer up a prescription about what one ought to do. The former are generally said to be the province of the sciences and some branches of philosophy, where the latter are sometimes understood as the province of ethics or politics.

Here's where it gets even trickier. It is easy to confuse the normative valence of descriptive claims with ulterior political motives, because prescription carries with it an ineluctable motivational and/or political component. Not only am I telling you what is, and not only am I implying what *ought* to be, but I'm very likely telling you *in order to motivate you* to act on what ought to be. This is true of all prescriptive claims, and it is also true about almost all ostensibly stripped-down descriptive statements and explanatory arguments. The normative oomph of those claims can't be cleanly separated from the motivational nudge or political ooh-la-la.

If a giant rock is about to fall on your head, and I tell you that I have observed this giant rock about to fall on your head, I am telling you in part (a) because I think you ought to get out of the way, and (b) because I want to motivate you to get out of the way. It is true, on one hand, that I am simply reporting the facts to you—"There's a rock about to fall on your head"—but it is also true that I am telling you for a *reason*—(α) it is bad for you, and (β) I want you to adopt this reason and get the hell out of the way. Whether that reason is narrowly prescriptive or widely motivational is a matter of great contention in philosophy. For now, however, just see how maddeningly intercalated the normative oomph and the motivational nudge of reasons are. Hold faith that we'll hopefully be able to gain clarity later.

Because of this conceptual entanglement between norms and facts, and then further between normative reasons and motivational reasons, scientific positions can never satisfactorily be extricated from the normative contexts and political arrangements in which they are uttered.[12] Most statements of scientific fact, no matter how descriptively pure and innocent they may seem, bear some prescriptive and motivational dimension, however faint. Much of the time we simply ignore this feature of scientific explanation because we more or less agree on the norms: "Gee, thanks for telling me that a rock was about to fall on my head. That would've been terrible."

Sometimes, however, the norms take on a much greater significance, and this is particularly true in tremendously important matters of public concern like climate change.

This peculiar feature of facts makes it all too easy to confuse our scientific forecasts—our statements about what "is" or what "will be"—with the reasons that we should stand up and pay attention to our environment. It's easy to think that we should be concerned about our actions because otherwise they will bring about massive climatic upheaval. Not unreasonable, of course. Climate upheaval would likely make for a pretty terrible day. But how so? Why? Would it be terrible for animals? For people? For future generations? The problem is that we just don't have clarity on what will be bad about it, or even whether it is bad in a way that should lead us to conclude that we ought to act differently. For this we need ethics.

DISASTER OF OUR OWN MAKING

What makes the climate issue so vexing is that it's not simply a case of rapidly rising carbon dioxide levels, nor is it a case of asteroids smashing fecklessly into the earth, nor is it even a case of diminishing terrestrial biotic matter; it's an impending disaster of our own making. With every gallon of gasoline we burn, with every cow we raise to slaughter, with every 3,000-square-foot single-family home we plop amid a field of perfectly manicured grass, we undo and radically transform the ecosystems on which we depend.

Most of us know that nature can send us horrific curveballs. Many of us seem unprepared to acknowledge, however, that we can also send ourselves horrific boomerangs. Whereas carbon emissions during the PETM were caused by natural events, the carbon emissions of today are partly the result of human actions. To be sure, there is a robust debate among climate scientists about the causes of climate variability. In a certain sense, they are trying to figure it all out before it's too late. Frequently, scientists and policy makers point the finger at carbon emissions, though most acknowledge that carbon dioxide is not the only driver of climate change.[13] The same scientists can go on for quite some time debating the merits of one or the other view. But this all seems to miss the point.

I'm here to report that the reasons you should be green cannot be found simply by extracting the normative valence from our scientific reports. If I tell you that a rock is about to fall on your head, it is relatively clear that you ought to move out of the way. But if I tell you that a rock is about to fall on your expensive car, it is far less clear that you ought to risk life and limb by moving the car out of the way. Maybe you are motivated to move it out of the way, but if you value your health and your life, you probably shouldn't risk it. So too if I warn you that a rock is about to fall on an innocent bystander. Maybe you have stronger reasons to move the rock in this latter case. What has to be in place is a moral reason—something telling you what you *should* do, despite your motivations, despite the politics, despite your unique and momentary desires.

And that's what makes the political emphasis on catastrophe ultimately problematic. It assumes (a) that all rational people will want to avoid a catastrophe of the sort that scientists are anticipating, and/or (b) that all rational people who may want to avoid this catastrophe will be motivated to do so. It assumes way too much about the target audience, which it does by smuggling in normative assumptions that masquerade under political and/or psychological (motivational) guises.

When we talk about climate change, we are talking about how the climate is changing, to be certain; and we are even talking about the causes of these changes; but more than anything, we're talking about what we're doing to ourselves. And when we compare the catastrophes of yesteryear with the catastrophes of today, we're also looking at an epochal demarcation that some theorists have called the *Anthropocene*: the period during which human activities have begun to have a significant influence on the earth's climate.

There is, therefore, one critical and glaring difference between the extinction of the dinosaurs and the anticipated outcomes from climate change. Where in one case extinction was advanced by some extraterrestrial influence or the churning guts of the earth, our current spate of global warming is believed to be primarily anthropogenic—which is to say, we're bringing this on ourselves.

Given that this is the distinguishing feature of what we're experiencing now, you'd think we'd spend a good deal more energy emphasizing the anthropogenic dimension and downplaying the catastrophic dimension. Instead we do exactly the opposite: we spend millions of hours and millions of dollars spelling out how bad it's going to be while spending almost no time or money on the question of what our responsibilities are. There's a reason for this emphasis on the science, of course. The forecasts are scary. Knowing what's coming out on the other side of the curtain might help us prepare a little better for what we face.

Nevertheless, if you're with me on the distinction between the descriptive and the prescriptive, then I think you must agree that just because some event is predicted to be catastrophic doesn't, in and of itself, require that we are obligated to respond to the expected catastrophe.

Don't buy it? Suppose, then, that changes in the climate will not be catastrophic, but instead will turn out to benefit a disproportionately large number of us. Some will suffer, let's say, but most will benefit. Suppose that a warmer climate will shift the agricultural base northward. Since much of the mineral-rich soil will no longer be covered in snow and ice, more food can be grown and more people will be able to eat. Suppose that the waters will be warmer, so that fish will flourish and pelicans will be healthy. Suppose that the warmer climate will mean greater precipitation in desertified zones, leading to better lives for wolves and bears, who may then spread and flourish as their habitat expands.

What then? Are we then to promote climate change? To drive our cars twice as far? Maybe we should engage in an enormous project of pumping all the earth's precious fossil fuel resources into an Iraq-sized basin and torching them. We might then more quickly arrive at our projected paradise.

It seems to me that even if climate change will, on balance, benefit us, it is not clear that our obligations to consider our fossil fuel consumption or our land use will change. I suspect, and I will argue over the next several chapters, that we are nevertheless required to consider our actions—our responsibilities to ourselves and others, as well as the impacts of what we do—even if, on balance, climate change will be good for us. What follows from such consideration of our actions, then, are real substantive obligations to address the various causes and impacts of climate change, including our emissions and our capacity to adapt. The difference is that instead of rooting this obligation in the impacts of climate change—the benefits and burdens, say—our obligations to address climate change arise instead from our civic obligations to one another.

The possibility of climate change benefiting us isn't entirely farfetched, after all. Even the mostly pessimistic IPCC fifth assessment report mentions that there may be benefits for some regions. Subarctic areas, for instance, may experience fewer deaths from cold exposure: "Positive effects are expected to include modest reductions in cold-related mortality and morbidity in some areas due to fewer cold extremes (*low confidence*), geographical shifts in food production (*medium confidence*), and reduced capacity of vectors to transmit some diseases."[14] And yet, though it is conceivable that climate change might be good for us, if it were projected to be so, I suspect you would still not hear calls from any community members to burn fossil fuels so as to bring about this projected utopia.

My attitude is that a climate catastrophe is only part of the moral problem, and that it's not clear that we should act any differently than if the climate weren't going to hell in a handbasket. Whether you will be helped or harmed by my stealing your car—maybe insurance will offer a handsome payout, say—doesn't change the moral upshot of my action. I've still stolen your car. Whether our current collective way of life will culminate, ultimately, in a devastating calamity or a fantastic paradise, we still have very strong reasons to change our behavior, to do the right environmental thing.

I'll say more about this in a moment. First I have to attend to one nagging bit of business.

PRELIMINARY EXPECTORATIONS: FORGET WHAT YOU KNOW

When I raise questions about the possibility of climate change being good for us—related such questions which I shall blithely raise throughout this book—I am not in any respect suggesting that climate change *will be* good for us. Quite the contrary. I am familiar with the science, and I know that climate change is likely not to be good for us, even with the requisite caveats about local variation and individual beneficiaries. Rather, when I ask these counterfactual questions, I ask only what your intuitions might be if the facts were slightly different.

Words like *suppose*, *if*, *consider*, *maybe*, and so on, should signal that I am trying my hand at tweaking your intuitions; that you should place the facts on hold, suspend your disbelief, and abandon the things you know. This is standard fare in my line of work. Philosophers, I often tell my students, don't traffic in facts. We traffic in arguments.

One might think it a deep problem to abandon facts, to turn toward the false or the absurd, but it's a pretty decent method, all told. Scientists fancy themselves to be doing exactly the opposite all the time. For good reason: if you want to be a good observer of facts, you would do well to put your ideological dispositions and your theoretical inclinations back on the rack. They interfere. They cause confirmation bias. If, by contrast, you want to be a good theoretician, you would do equally well to put your observational apparatus in the closet—to shut out the light and focus instead on the way our ideas hang together.

There will be times throughout this book when I imagine states of affairs for which there is absolutely no supporting scientific evidence. I will ask that you imagine absurd things, such as that climate change may be good for us, that you are asked to nibble on your own hand for nourishment, that you are to wear aluminum foil underwear during a lightning storm, that we can build a global thermostat to control our climate, or that we can inject a health potion into our water supply. When I do this, I want to call attention to the argument, not to the truth or falsity of the absurd premises, and not to the ultimate conclusion. When I raise an *if* claim, what I expect you to do is to evaluate that claim, operating under the assumption that what I say is true even when you know that it is not true. The appropriate response is not to object that you would never, for fashion or comfort reasons, wear aluminum foil underwear. We're not playing the same game if you don't mentally don my uncomfortable knickers.

This is all very tricky going—more than you might first assume. Indeed, there are times that we deploy our *if* claims indicatively, so the facts really do matter. If there is milk in the refrigerator, and I wasn't the one who put it there, then someone else must have done so. This is a very comfortable way of making an argument. But it's an *if* argument of a different sort from the type that I will ask you to consider here. It would be easy to misunderstand my objectives if you think of my arguments in the indicative: "If climate change is good for us, he's saying, then we should do what we can to burn our gasoline now. But climate change is not good for us, so he is being stupid."

Put a little differently, I am not stoking denial by asking an *indicative conditional*, as Fox News commentator and former Nixon speechwriter Ben Stein asked, as so many other climate skeptics have asked: "What if climate change is a fraud?"[15] I am not being provocative, like columnist George Will, who suggests that perhaps the benefits of allowing global warming to proceed will outweigh the costs.[16] Maybe the world will be better for agriculture, he opines, or perhaps it'll be a beachy paradise.[17] No. These commentators are sowing the seeds of doubt to score a political point.

Rather, I am raising a *counterfactual conditional* to make a considerably more conceptual point about the reasons we should be environmentalists. Imagine, contrary to the facts, that climate change will be good for us.

My argument trades in modalities, in possibilities; and not just any possibilities: logical possibilities. It stems directly from my first two observations: the first, that I am making an argument, not reporting the data; and the second, that I am trying to piece together how things ought to be, not how they are. How the world ought to be, I assume, should apply no matter what the current state of the world is. By assuming counterfactual conditions, we can test the stability of our conclusions *even if* some false claim holds.

Here you have it, then: an environmentalist who is a proud member of the surreality-based community. Permit me, for this work, the luxury of titillating your inferential substructure. Relish this. It's a rare treat to read a nonfiction work in which the author proudly decries reality.

CATASTROPHE REDUX

At about the same time that the dinosaur bones of the Very Reverend Dr. William Buckland et cetera et cetera were traveling through the circles of London and Gideon Mantell was regaling the scientific establishment for its reliance on the explanatory numbskullery of the Old Testament, the French naturalist Jean Léopold Nicolas Frédéric Cuvier, known to most as Georges—for reasons that utterly escape me, given that he had four perfectly fancy and acceptable other first names—gave voice to a pessimistic theory about the chronology of the universe, colorfully dubbed *catastrophism*.

Catastrophism proposed that the fossil bones discovered by Anning, Buckland, Mantell, and others not only were considerably older than first assumed—by "thousands of centuries"— but also were so dramatically unfamiliar because the world has developed not slowly and gradually, but in a god-forsaken chain of miserable geological upheavals. Catastrophism thus stood in sharp contrast to *uniformitarianism*, which until Cuvier's time was the prevailing view. Naturally, uniformitarianism proposed that the laws of nature are uniform, that the world that goes behind the curtain doesn't do anything spectacular before it reemerges on the other side of the curtain. It proposes that physical laws stay the same, that geology doesn't shift in crazy ups and downs.

Like many fashionable theories, Cuvier's version of catastrophism was too radical to last long. It went out of vogue not long after it was proposed, and uniformitarianism once again became the dominant presumption about the history of the universe. This dominance lasted for a full 150 years, until one day in 1980 when a paper by a father-and-son team, Luis and Walter Alvarez, exploded out of *Science* magazine, hypothesizing that the aforementioned killer asteroid had shattered the earth's equilibrium and given rise to the Cretaceous-Tertiary extinction, or the K-T extinction, which most of the rest of us know as the day that the dinosaurs met their collective maker.[18]

At about the time, in other words, that Ronald Reagan took office in a landslide election and began arming the nation to the teeth with nuclear weapons and fantasy laser beams, thereby pushing the United States ever closer to the teetering edge of a nuclear war with a sleeping bear—ensuring, in effect, that catastrophe would be mutually assured—catastrophism came right back into vogue. The paleontological establishment began to accept the hypothesis that a tremendous asteroid had come hurtling through the cosmos to smash all earthly critters to smithereens. Like a bomb.

Also about that time, our fair weather fellows at the National Center for Atmospheric Research, as well as numerous other scientific outposts, including NASA, the National Oceanic and Atmospheric Administration (NOAA), the Woods Hole Oceanographic Institute, the Climatic Research Unit at the University of East Anglia, and so on, began piecing together a somewhat different puzzle about the state of the climate. By digging deep into glaciers to extract ice cores, by reconstructing temperature records from centuries of tree-ring data, by examining the strata of the earth, the climate establishment slowly began to accept the incredibly daunting possibility that, just by living our otherwise unobtrusive lives, we were opening the doors to a real, nonhypothetical climate catastrophe.

The planets aligned, the politics converged, the scientific studies coalesced, and enthusiasm for catastrophism was renewed. None of which is to say that the threat of climate catastrophe is not real, that it is somehow a political fabrication. I trust that the threat is real, that the thousands of climate scientists who research and write about this topic are slowly uncovering a terrifying phenomenon. But like most people, I am not a climate scientist. I'm not qualified to evaluate the science. Instead I rely on the scientific community to help me understand what's going on.

The important point, however, is not the truth or falsity of climate change. It is that catastrophism, as I've said, trades on a radically different picture of nature than uniformitarianism; it trades on a picture of nature as wild and destructive, as a force to be reckoned with—a force that will not only ruin your morning, but ruin *all* mornings, which seems to be a very big deal, even for people who don't much care for mornings.

I suspect that those who shout from the rooftops about the impending climate catastrophe do so in part because they think that catastrophe is an incredibly powerful motivator. The end of nature is nothing to take lightly, and the threat of catastrophe has been employed to great effect by those who either are genuinely concerned about the eventual destruction of the earth or would at least like to see people change their behavior. For those of us with children, catastrophe can offer double-barreled motivation. I certainly don't want to leave my son a barely habitable world. I'd like to have him enjoy the benefits of nature that I've enjoyed. I'd like him to see the things I have seen and even be able to pass these things on to his children. If ruin truly is the destination toward which all men rush, as Garrett Hardin once opined, then maybe a hard kick in the ass is exactly what we need.[19]

The problem for environmentalists is that this is a really complicated political gambit. I'll say more about the state of the environmental movement and environmental theory in the next chapter. For now, since statements of fact are wrapped up in politics, all of which are shot through with values, it will help to discuss the significant downside *political* risks of hanging our hats on climate catastrophe, which will help make my case for an ethical approach to this problem. As I am talking here primarily about ethics, politics are not my main concern, but the political risks make for a good stepping-off point. If we can get the ethics right, then I'm hopeful that the politics will fall into place.

First and foremost, terrifying long-term forecasts are not a motivator for everyone, and for a good number of people they simply do not resonate. Who cares if humanity dies out, or if the earth is irrevocably altered? Most of us won't be around to witness the really terrible stuff, so climate change is primarily a concern for future generations. You've heard this kind of selfish response many times before: "The real catastrophe won't occur for another 100 years? Then what concern is it of ours?" (Some scientists, partly in an attempt to counter the indifference illustrated here, have tried to show the near-term effects of climate change.)

Moreover, emphasizing catastrophe can be extremely divisive.[20] This talk of catastrophe raises the hackles of those who have investments and interests in the status quo. Naomi Oreskes and Erik Conway observe that full industries are in place to sow the seeds of doubt.[21] They draw parallels between the climate contrarian community and those who historically battled the scientific research on the dangers of tobacco and DDT, suggesting that small niches of the scientific community, coupled with a media eager to portray multiple sides of a story, work collectively to sow the seeds of doubt in the minds of the public.

Further, it would be very easy to characterize this emphasis on catastrophe as a simple matter of framing. Some academics familiar with the political scene, like George Lakoff, add fuel to the fire by pinning responsibility for the political sputtering of climate change on the "characterization" of climate change.[22] Critics typically angle at the motivational impotence of focusing on the negative. They suggest that rather than fighting fire with fire, a better response may be to change the frame of the debate entirely, to focus on the energy discussion (say) or on a new jobs bill. They encourage us to focus on the "aspirational view," to look to promote alternative energies, or to build a new energy economy. This was the response from Obama administration officials such as Rahm Emmanuel and David Plouffe. But the problem runs quite a bit deeper than that.

It's not that I think there are no cultural issues that may be guiding the discussion, or that political considerations are irrelevant. Surely there are. But many figures in the climate discussion rely so heavily on these political questions because this is likely the only prescriptive forum they know. Something must be done, they insist. That seems clear to them from the science. The arena for discussion of what to do is the political, so that is where they end up. All the while they ignore the ethical discussion.

When we slip into frames language, we make a serious category mistake. Whereas the strategic communication discussion may appear to be prescriptive, in the sense that depending on how the climate problems are framed and communicated to the public, people are likely to change their behavior accordingly, it is not prescriptive in the sense that it drills down into the question of what it is *right* to do, identifying what we ought to do and why we ought to do it. In the end, I feel confident that the strategy discussion is not what will move people; but instead the ethical discussion.

It doesn't matter to our moral obligations how we frame an issue. It doesn't matter, for instance, to the rightness or wrongness of abortion if we speak of it in political terms, as a choice, as a life, or as a law. Even if climatic changes don't reach the level of a catastrophe, and no matter whether we frame the impending climate changes as *global warming*, *climatic change*, *climate variability*, or an "endless, surf-happy summer," it seems to me, we are still obligated to do something about climate change. We are still obligated to reduce our emissions and our greenhouse gas footprint, even if our actions make the world demonstrably better for a large number of people. Whether the effects of climate change are calamitous or fortuitous has little bearing on how we should act. Whether environmentalists frame it as such has little bearing, too.

If you buy my argument above that it doesn't matter whether climate change will be good or bad for us, then you have taken the first step to seeing the point that I advance throughout this book: that we should take responsibility for our actions, and that we should act responsibly and respect others, even if our actions will not result in catastrophe, or bad outcomes, or in some cases, make the world better.

THE TERRIBLE, HORRIBLE, NO GOOD, VERY BAD FUTURE ARGUMENT

What I'm talking about above primarily relates to the question of framing, which is fundamentally a political matter. I've suggested that the catastrophe frame isn't clearly a conduit to effective political action. Part of the reason for this may have something to do with human nature—maybe we're just not motivated to act in the face of catastrophe—but some of it may also relate to complications stemming from the logic of catastrophe thinking, or what we might call *catastrophe logic*. Indeed, if we tease these complications out, we can see that there are much deeper, much more problematic *ethical* objections to catastrophe logic. To see why the threat of catastrophe will not do the ethical work that some people hope it will do, it will help to look a little more closely at the reasoning that underlies it.

One problem with catastrophe logic is that it's not particularly sound. There is an enormous range of catastrophes that we would do well to avoid: asteroids, plagues, earthquakes, and climate change are just a few. According to one way of looking at things, we should make our decisions about what to do based on a catastrophe's likelihood and its projected damage. Given the magnitude of the catastrophe, as well as the probability of that outcome, it would make the most sense to avoid the most calamitous catastrophe. We can calculate an expected value using this formula.

It doesn't take much manipulation of the numbers to see that this is a losing prospect. A well-targeted asteroid would completely annihilate our planet. There would be nothing left. No matter how you slice it, this is a very bad outcome for the earth. It doesn't get much worse. If given the option between an asteroid strike and some lesser catastrophe, it almost certainly makes sense that we would choose the lesser catastrophe. I suspect, for instance, that most of us would choose a worldwide epidemic of Ebola over an asteroid strike. Fortunately, the projected likelihood of a cataclysmic asteroid strike during our lifetime is vanishingly small, estimated to occur only once every 600,000 years.[23]

Before you go feeling better about the prospects of such a thing, remember that we're looking at avoidance of *catastrophe* here. If we calculate it out, what we're talking about is a relatively high likelihood of an Ebola pandemic against the very small possibility that a measly little asteroid will come along and ruin everything. Given that the devastation of an asteroid strike is extremely terrible—maybe even thousands of times more terrible than an Ebola pandemic, which is pretty darn terrible—it would be better to avoid the asteroid than to fight Ebola. According to catastrophe logic, we should dump all of our resources into preventing an asteroid strike, *even though such an event is extremely unlikely*.

Another problem with catastrophe logic is that catastrophes loom around every corner. There is a small but nonnegligible chance that you will die in a plane crash tomorrow or that you will be clunked on the noggin by a falling piece of space garbage. If you subscribe to catastrophe logic, you ought to do everything in your power to avoid getting on that plane or walking down the street when satellites are overhead. Catastrophe logic appears to put us in a position where we can't do much of anything. It induces paralysis. Since any given action could result in innumerable awful catastrophes, if we were to heed catastrophe logic, we'd likely be wrapping our children in bubble wrap and shutting the doors.

Further, all this talk of catastrophe ramps up the apparent permissibility of any and all evasive measures. If the outcomes of inaction are going to be terrible, then we really had better get our act together in order to preempt the terribleness that is likely to ensue. With regard to other people, it plays the "ticking time bomb" card. If we take catastrophe as a given, it is outrageously easy to justify almost all horrific behaviors.

Want to avoid a catastrophe? You should do whatever it takes to do so, even if it involves trampling the rights of a few—or in the case of cataclysmic disaster, maybe even the rights of many. Nuclear device primed to detonate over New York City? Round up the suspects. Any who are suspected of knowing how to defuse the bomb, in whatever way—if they've even crossed paths with those who might—and all bets are off. Torture away. Population got you down? Nothing a few wars can't solve.

"Ticking time bomb" scenarios effectively halt rational discussion about what to do by stanching any considerations that do not contribute to averting catastrophe. In effect, they put a gun to our collective head. The more we appeal to catastrophe to justify our actions, the more likely we are to permit otherwise morally problematic interventions.

I think that environmentalists can take a different, and perhaps a better, route to get to the conclusion that we should do something about the climate and ultimately about the environment. We can arrive at this conclusion without spending all of our time and money trying to persuade one another that our science is right. For the purposes of deciding how to act, we don't need to know whether the climatic changes are going to be 2 degrees, 3 degrees, or 40 degrees. For answers to those questions, we need to look elsewhere. We need to think differently about what we do and why we do it.

2

The Precious Vase

In which the author (1) introduces the reader to the components of an environmental action, (2) argues that the current emphasis on climate builds on the same old environmental story, (3) presents the "precious vase" view by way of two approaches to environmental value: the intrinsic and the extrinsic, (4) briefly explores several holistic theories of intrinsic value, which he (5) chases with individualistic theories of extrinsic value only to (6) outline the pitfalls of sticking with the aforementioned precious vase view.

LOCKDOWN

Only moments before Bobcat was to hoist his body and his gear 30 feet up onto the platform over the dusty road where he had been suspended for the previous 12 hours, Jaguar sent out a warning call. "Doughnut convention," his voice crackled over the radio. I snapped to attention. Jaguar, or so he anointed himself, was positioned on the other side of the mountain with binoculars and a walkie-talkie. I ran up the hill to notify the others. Fox, Squirrel, Spoon, and Spartacus were waiting at their stations. Otter had dozed off and buried his head in his sleeping bag. Clam and Oyster were cleaning pots by the fire. Tens of other forest critters were scattered about the face of the mountain, snoring peacefully as dawn's rosy fingers tickled the rocks and wisps of steam danced on the warming grass.[1]

"Doughnut convention!" I huffed into the radio, running up the mountain, partly gasping for breath in the thin mountain air, partly freaking out with knowledge that we would soon be set upon by law enforcement.

This otherwise laggard and dirty group of hippies, tired from long nights sleeping on the hard Colorado ground, exhausted from late hours talking politics and strategy around the cookstoves, spun into action. Squirrel ran to his post, Timber to hers, and I back to mine, grabbing a pen, paper, and my camera. Sleeping bags were shucked, hats were donned, water bottles were filled, food was shoved into mouths, and U-locks were redistributed to those who earlier had volunteered to put their necks on the line to save the trees.

After this momentary flurry of activity—silence. The slow scratching of boots on pebbles amplified the tension. And then? A pause, before the site would explode with activity.

What ensued was a Keystone Cops routine that would make Charlie Chaplin proud. A slow chant came from over the crest, and before we knew it, approximately 25 officers of the law, all wearing black or blue flak jackets, surrounded us, muttering meaningless "Hut, hut, huts." It was like that ridiculous scene in the *Blues Brothers* in which hundreds of G-men scale the walls of Chicago City Hall, only to trip and stumble over one another in the growing chaos. I was immediately overtaken by a sudden and uncontrollable case of the giggles.

• • •

Forest defense actions are nothing if not outrageous. When recounted in the company of the uninitiated, they play out like the wartime mash-ups of Aesop's fables and the Battle of the Bulge. The cartoonish monikers—Bobcat, Jaguar, Fox, Squirrel, Spoon, Spartacus—are unique to every activist, sometimes unique to a particular action. Fickle like the breeze, they can change with the day, much like the people who inhabit them may hop from campaign to campaign. Sometimes the change comes because the activist needs a new cover. Sometimes the change comes simply because the consequences of naming oneself are not immediately apparent. My friend Fire didn't realize the shortcomings of her chosen name until it was shouted during a particularly tense time, which worsened rather than aided our panic. A crowded theater it wasn't, but hers was not a good name to shout under even placid circumstances.

Forest names are chosen for all sorts of reasons, and many are tongue-in-cheek, cynically attuned to the absurdity of the protest situation. Ridiculous though the situations may be, direct actions such as these are not the flaccid and toothless sit-ins that are sometimes depicted in caricature. They are highly orchestrated affairs, involving often extraordinary planning, coordination among disparate groups, travel over long distances, and boot camp–style training sessions long before any direct action actually occurs. There are clusters of social justice organizations subbed out to every possible need, each ready to pick up at a moment's notice.

Organizations like the Center for Biological Diversity and Wild Earth Guardians conduct extensive biological surveys and initiate legal proceedings against any and all environmental villains. Loose-knit political action groups like 350.org, Ancient Forest Rescue, Beehive, and the Rainforest Action Network head up political campaigns to publicize and get out the message. Culinary collectives like Food Not Bombs and Seeds of Peace roll out their caravan of jalopies and cookstoves to feed the troops. The Student Environmental Action Coalition works on campuses to channel the energy of young students into the campaign. The Ruckus Society gears up to train the neophytes in the rudiments of climbing, banner hanging, media relations, lockdowns, tripods, and so on. The Sea Shepherd Conservation Society, Greenpeace, and Earth First! gather to throw a monkey wrench into the gears, just before an event of environmental destruction is about to go down. Law collectives arrive to give activists the inside lowdown on their rights and to defend them if they get arrested. In many cases this motley array is convened to satisfy a seemingly straightforward objective: to get the government to enforce its own laws against well-funded private-sector actors who have the resources and the wherewithal to know how to skirt the law in their favor.

This division of labor extends even throughout the action itself. Though not a general rule, direct actions tend to work best if all participants have defined roles. For starters, there are "arrestables" and their handlers: those who either intend to be arrested or are in place to ensure that the arrestables are not physically attacked by loggers and are treated fairly by police. There is generally a press team, equipped with press packets and explanations of what's up. There are conflict mediators positioned to soothe aggression as it wells up on either side. There are legal scribes, documentarians, photographers, nonarrestable support, and so on. For every activist there is a role.

Many of the activists are well-bred and well trained, sometimes coming from positions of surprising authority, education, knowledge, and privilege. There are field biologists, forest ecologists, lawyers, geophysicists, and in more than a few cases, starry-eyed philosophers. Most are extremely articulate and politically astute. Almost all are frustrated with the current state of affairs, and almost all are worried that we are destroying the earth.

Funding is a different issue altogether. Forest actions are typically pulled off on a shoestring budget. Supplies are often acquired by passing the hat. Dumpster divers head out on nightly scavenges to retrieve discarded meats and other goodies from the metal clamboxes behind restaurants and grocery stores. More than once I have witnessed activists busking—juggling, drumming, fire-breathing, panhandling—in the streets to drum up money to fund the next action. Indeed, despite my musical inabilities, I myself have busked with drums and fire and juggling batons on the streets to gain support for and interest in our campaigns.

What is most impressive, however, is that forest actions actually yield results. At a minimum, the spectacle of the protest gives otherwise oblivious and disengaged observers—those who, say, read about it later in the paper or watch it on the news—a thing or two to talk about as they watch it all unfold. They may not have the nicest things to say, but at least they say something. If the activists make news, they may inspire a discussion or two around the dinner table. And media attention also raises the profile of the environmental issue in the eyes of the courts and the public.

Captain Paul Watson of the Sea Shepherd Conservation Society was widely criticized as an opportunist, a sellout, and a jerk for taking Sea Shepherd's actions public when he signed a contract with the television channel Animal Planet. And yet, his was one of the more effective tactical moves in recent memory. Sea Shepherd, which had otherwise been a relatively lone wolf in the war to save the whales, is now gaining momentum as it attracts publicity. Watson explains as much on the show, but goes into greater detail in his book *Earthforce!: An Eco-Warrior's Guide to Strategy*, portions of which I assign to my undergraduates.

The same can be said for smaller but no less effective organizations. The Center for Biological Diversity, for instance, has had tremendous success prosecuting violators of environmental laws. While the cases themselves have legal merit and are often won in court, in many instances the center's success has been secured in part through the publicity of direct action. Almost everyone in environmental politics will tell you that radical organizations can apply critical pressure to move policy forward.

Though forest actions still play an integral role in the environmental movement, they are not nearly as often encountered now as they were during the 1980s and 1990s. They were, in a way, a product of the time, when environmentalism remained relatively fringe. It has since matured. It has come out of the woods and taken the world by storm, largely as a result of increased concern over the climate. This has been both good and bad for the American environmental movement. On the upside, what could be better? Where before the concern about the environment was limited to a select few, now millions of Americans, and millions of people worldwide, self-identify as concerned about the environment.[2] On the downside, it is not at all clear what it means to self-identify as "concerned about the environment," and there is even reason to believe that concerns may be misplaced.

CANCELED DUE TO WEATHER

Anybody who has ever had occasion to stumble around the darkened campsites of an environmental action just before daybreak, or who has ever been so brazen as to voyeuristically but disinterestedly shuffle through a large-scale environment and social justice march, or who has spent any time at all around the legions of fledgling environmentalists on college campuses, will know that what appears at first blush to be a monolithic environmental community is, in fact, an incredibly diverse collection of ragtag nature lovers and antiestablishment types—a cook's stew of NIMBY activists and crystal worshippers, geeked-up technophiles and dressed-down Luddites, milquetoast recycling junkies and malodorous cavemen, all thrown together and peppered with an indeterminate number of crazy people.

The "crazy-people factor" is true about almost all social movements, of course. In any given demonstration you're bound to find misspelled signs and cockamamy slogans paraded by illiterate nincompoops, much to the delight of the opposition. Right-leaners and left-leaners alike enjoy a healthy serving of nut jobs. It is the nut jobs who typically make the news, but it's not only nut jobs, of course. The ideological wings of Fox News and MSNBC, as well as the meta-ideological cynics at *The Daily Show*, make great hay of the inconceivably stupid things that normal people say.

Anybody who has observed this diversity in the environmental movement will no doubt also be aware of the extent to which the environmental discourse has in recent years tilted strongly in the direction of climate change. Think back just a decade or two ago. Environmentalists were soundly ridiculed for wistfully pleading to save the whales, the spotted owl, the snail darter, and the jaguar. Flip now to almost any environmental organization—or heck, head to your local supermarket or home furnishings store—and you will notice, no doubt, that the vast majority of literature associated with what is "green" is focused primarily on the climate. There is scant mention of animals, almost no recognition of trees, widespread disregard for water, little talk of mountains, not to mention nuclear proliferation, midnight toxic dumping, environmental racism, biodiversity, pesticides, or tribal rights—unless, that is, the vestigial organs representing these issues have succeeded in finding some way to rope themselves into our unique Anthropocene conundrum. Environmentalists of every stripe are either hitching their wagons to climate change or floundering about, trying to regain the public's attention.

The recent UN Conference of Parties (COP) meetings (e.g., in Kyoto, Bali, Poznan, Copenhagen, Cancún, Durban, Doha, Lima, and Paris, to name just a few) stand as testament to this plurality as well as to the extent to which the political arms of the environmental movement have shifted their strategy to capture the political capital currently dumped into the climate issue. Conservationists, who had otherwise been concerned with broader issues regarding biodiversity and the protection of endangered species now focus primarily on the plight of the polar bears and other token megafauna.[3] The antinuke folks have folded up their placards, packed away their Geiger counters, and reiterated their interest in alternative energies.[4] The antipollution activists have moved away from nitrogen and sulphur oxides emissions (more affectionately known as NO_x and SO_x emissions) and toward carbon and greenhouse gas (GHG) pollutants.[5]

One can dispute this, of course. It's not as if there aren't other green concerns that seem to be gaining momentum. The local-food movement, for instance, doesn't seem inordinately infected with the climate bug. But even some of the motivation for this movement can be roped back to concerns over climate change. Refer to "food miles," for instance, and you're talking mostly about the climate.

Evidence that these arguments are fraying at the edges is ubiquitous. With almost every sneeze or cough of the climate system, there are charges on both sides of the debate that that particular weather event is validation that climate science is either right or wrong. When there is a deadly heat wave or drought, you can count on the guardians of climate politics to proclaim how this is "consistent with" expectations.[6] If there is a record snowstorm, or if the winter rolls along without interruption, leaving people to feel as though it is considerably colder than normal, then it is not long before one of the blaring sirens of the political right-wing blogosphere points out the alleged irony that politicians are meeting to discuss solutions to climate change.[7]

As it happens, many people in the climate change community express frustration that these facile weather claims are used by partisans to advance a political agenda. To counterpunch, they patiently explain that weather is not climate, that the statistical significance of any given weather event is almost never high enough to say that that specific event is caused by or correlates with the models.[8] Saying so is like saying that some particular sneeze or cough was the direct result of seasonal allergies. "One swallow does not a spring make," as Aristotle says.[9] They will then wring their hands at the misinformed public. Some scientists even distribute talking points and "tips for dealing with the media" to try to refine or simplify their message so that the public can understand the science.[10] Environmentally concerned critics of the climate community typically warn that using hyperbole to communicate a message actually opens the climate science to greater criticism. Al Gore was taken to the mats for his use (or abuse) of hurricane disaster information to drive home the climate point. Meanwhile, climate bloggers like Joe Romm rip fellow environmentalists to pieces for raising concerns about the disintegration of the broader environmental movement. This is a point of incredible tension.

The political pressures to fixate on the climate are enormous. To my mind, this is a lamentable error, rooted not in bald opportunism or political miscalculation, as one might assume, but mostly in a mistaken view about why we should be green. Although climate appears to be a new direction for environmentalism, it's nothing new. Environmentalists have for decades been calling to protect nature, to keep it from being sullied or damaged. Substitute the word *climate* for *nature* and you have your parallels. We've just shifted our focus away from protecting nature in general to protecting the climate, to keep *it* from being thrown out of balance. This emphasis, of course, is not entirely unreasonable or disingenuous—if the climate collapses, there too goes nature. Meanwhile, most every other environmental issue is summarily swept under the rug.[11]

But none of this is my concern. As I've said, I want to put politics to the side. I spent some time in the previous chapter discussing the political implications of the catastrophe frame and then chased it with a brief discussion of catastrophe logic, hoping there to show how the logic can underlie the frame. It may help if I reiterate here that, as an ethicist, my central concern is not with the strategic implications of any view about nature—I don't really care in this book how the situation might unravel politically (though I care very deeply about this as a person living on the planet)—but rather with what's right, with having the right reasons. I suspect, as a matter of fact, just as I suspected with catastrophe logic, that the above political concerns are a symptom of the underlying problem: we've got our reasoning wrong.

As it happens, I consider myself an environmentalist primarily because I think environmentalism is right. It strikes me that the main reason to be an environmentalist has nothing to do with climate or catastrophe. Again, my attitude here is neither because I am skeptical about the climate science—I'm not—nor because I think that emphasizing climate is a bad political decision, but because both climate and catastrophe are the wrong reasons to be an environmentalist. Further, I suspect that many people instinctively sense the lacuna here, and that this may be the reason that there are substantial downside political risks of hanging our hats on climate catastrophe. So it will help, then, to stir up this incredibly diverse pot of stew.

TREE HUGGER 101: THE BIRDS, THE BEES, AND THE LOVELY TREES

When I was in college my mother gave me a small framed picture of the earth inscribed with LOVE YOUR MOTHER. Since then I've lived in perhaps 10 different dwellings, and yet I still have that trashy piece of eco-kitsch. I can't say why. It sits at the bottom of a box, underneath a hundred other scattered mementos of an earlier period in my life. It has never hung on a wall.

Like many who have a healthy appreciation for nature, you might inexorably be drawn to the view that you should love the earth. Perhaps, like many, you have come to believe that our unhealthy relationship to nature is characterized by our historical disposition to mow down and steamroll whatever threatens us. Aggression is best countered with love, you might assume, and so it is love for or appreciation of nature that is in need of cultivation. Having made this logical connection, you might further conclude that the best way to advance this view is to demonstrate that nature is not something to fear or abhor, that it is not bad for you or bad for others. You might, that is, try to demonstrate that nature is, on the one hand, extremely valuable, but on the other hand, extremely fragile—like a vase. And you might, as a consequence of this view, think that nature must be protected, or defended, or preserved. Your thinking might look something like this:

1. Nature has value (for such-and-such a reason).
2. We should promote (or protect) things of value.

---

3. Therefore, we should promote (or protect) nature.

In the mid-1990s, frustrated with graduate school and angered by the slow progress of the environmental establishment, I certainly adopted this attitude. More than once, I recall breathlessly exclaiming just how grand it all was. Meanwhile, I would be fretting over my wind-chapped face or the hazardous cliff at my feet. In the back of my mind, I felt a little uneasy at getting doe-eyed over deer or orgasmic about rainbows.

It's not that I didn't mean what I said. I did—or, at least, I thought I did. It's just that it wasn't always clear to me that nature was as magnificent as I was then saying. What I think was going on was that I was responding to a particular kind of argument. Intellectually, I had grasped that I ought to do something about the reckless and wanton destruction of forests and streams and deserts. It seemed to make sense that the reason that I should do something about this reckless destruction was that there was something in nature to value. So, practically, as one who did not naturally see this value, I tried to train myself to see it by cultivating the right reaction. I think I probably did the same with regard to other things that I thought I should value: the opera, ballet, trance music, conceptual potty art, the dilapidated and dangerous housing that masquerades as historical preservation, and so on.

As a consequence of my intellectualized moral position, I further decided that I wasn't doing enough to help bring about environmental change. I spent several summers in a row, and many a cold weekend, out in the real world, blockading roads, planning lockdowns and direct actions, confronting loggers and police, being arrested, and spending time in jail.

A *lockdown*, if it isn't entirely clear, is precisely what it sounds like: a small brigade of concerned citizens who will lock themselves to each other or to a gate or a barrel in the road in order to protect the earth—a forest or a stream or even a mountain—before man's implements of destruction can tear the place to shreds. A lockdown is also, incidentally, not unlike what we do with our very valuable objects, like priceless and precious vases. We put them in safe places or safety deposit boxes.

The "precious vase" view, or the idea that nature carries some sort of value (and is therefore in need of protection or promotion), can be advanced along at least two lines, among which there are a plethora of possible argumentative strategies. One can either try to advance the position that nature maintains *intrinsic* value, meaning that there is something valuable in nature itself, apart from external considerations. Or one can try to advance the position that the value of nature is primarily *extrinsic*, meaning that the source of its value lies elsewhere. If you think about it, it's not unlike the classic optical illusion below.

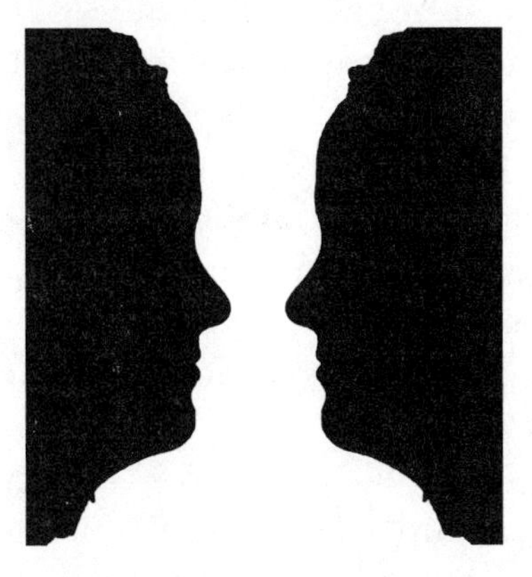

The intrinsic value view is reasonably common. It's the idea that trees or rocks or sometimes even species and ecosystems have value in themselves; that even in a world completely bereft of human beings—a world in which, say, we were just picked up and whisked away—there would still be some value in those objects. Look at the vase. That's where the value is.

The extrinsic value view is also reasonably common. It proposes that trees or rocks or sometimes even species and ecosystems have value because of some other factor. Often, the concern is that trees or rocks or sometimes even species and ecosystems are valued by other people, but the value could equally well come from other sources. Look at the faces. That's where the value is.[12]

There really is a range of supporting arguments that one can take with regard to the intrinsic value of nature. Many contemporary environmental philosophers think that by isolating value in some class of individuals, they can identify value in wider classes of individuals. These *individualists* (as opposed to *holists*) may find value in a unique attribute, like life or subjectivity or the capacity to have interests, say. If we value such attributes, goes the reasoning, then we should value those same attributes wherever we encounter them. Higher mammals are the most obvious candidates as bearers of value, so the discussion usually starts by isolating attributes of these mammals.[13]

Others take a religious view and propose that God has endowed the world with certain value that we are to steward and protect. Still others take a more romantic approach. Maybe they've dutifully read their Wordsworth, Shelley, Thoreau, Muir, and Snyder and find it compellingly clear that nature is magnificent and glorious. Some just think that there's something more there, something that isn't captured by our relentless desire to trap and contain nature, to monetize nature, to put a price on the earth. There are a lot of good reasons to be intrigued by the notion that placing a dollar value on the natural world is nutty.

To be straight, I have my sympathies with some of these positions, but I don't want to talk about them here. Rather, I'd prefer simply to note the incredible diversity of argument defending this view that nature has intrinsic value. I'd also like to show briefly how riddled with complexity these arguments can get.

It takes a certain sort of academic abstraction to isolate specific attributes of nature, or things in nature, and then to observe that all entities that have those attributes therefore have value. Many people are naturally turned off by this approach. Nevertheless, this is one of the more prevalent strategies in environmental philosophy. The important point is that these arguments attempt to isolate the value of nature or of entities in nature, and they do so largely as part of a longer argument aimed at getting us to change our behavior.

I'll just make one quick further observation. When you talk about the intrinsic value of nature you damn well better be clear on what you mean by *nature*, which is not at all to raise the other complicated metaethical questions of what we mean by *intrinsic* and what we mean by *value*.

Consider just a few of the ways we deploy this confusing term. Sometimes we use the term *nature* geographically, to refer to the untouched and the untrammeled: areas of wilderness lying far outside our city walls. Sometimes we use the term *nature* cosmologically, to refer to "great chains of being," in which we human beings are positioned just beneath the gods. Alternatively, we break those chains and introduce new non-hierarchical orders to nature: I am my mother's son, and she her mother's daughter, and her mother her mother's daughter and so on down the line, back to the apes, who are the daughters and sons of other creatures, only remotely like them, which are, then, the primordial backwash of an electrical storm in pre-pre-history. Sometimes we use the term *nature* to refer to instinctual nature, or sometimes even *human nature*. When we use it this way, we imply that there is some teleology, some "purpose," some way that "things are" that explains the actions or behaviors of entities in nature. We intone, channeling David Attenborough: "it's in the Great Grizzly's nature to seek berries and salmon," just as much as it's in our nature to seek a mate and to respond aggressively when our chances of procreating are diminished. Sometimes we use the term *nature* in a more Darwinian way, to refer to a chaotic proving ground on which only the fittest survive, in which life is a struggle for genetic domination and persistence through time.[14] We sometimes use the term less metaphysically, suggesting that nature is anything that has been untouched by humans. So we might call *nature* the nonhuman or the artificial. Any X that satisfies the condition that it has not been touched by humans, we might call *nature*.

These aren't the only views on nature, but they are some pretty common views. The question of what is natural appears again and again in philosophy journals. It is less frequently encountered in popular culture, though it nevertheless appears. It makes its strongest appearance on grocery store shelves, where we can see the results of the US Department of Agriculture's extensive debates about what can be called *organic* or *natural* and what cannot. All chickens sold in grocery stores are said to be natural, even though they are raised in cages and processed on an assembly line, like tube socks or cans of spray cheese. The question of what is natural can be exceptionally abstract and philosophical, but it is pretty important to a lot of people.

I can't go into a pages long examination of the various uses of *nature* here. That's a topic for another book, and I actually think they are questions we will gain insight into as we move further. For now, I'd prefer simply to bracket these questions and refer to nature colloquially, as the nonhuman, the other-than-human, the nonrational: it is the crashing of a waterfall, the clap of thunder, the movement of the earth, the sky, the birds, the bees, and the trees. It is a force that surrounds us and a force that is in us; and yet, it is distinctly nonrational. It is this, I think, that environmentalists are concerned to protect.

CANARIES, COAL MINES, AND THE WEB OF RELATIONS

When Julia Butterfly Hill sat atop her redwood tree defecating into plastic baggies and sending her waste down a line to handlers on the ground, she embodied an attitude of connection to nature that makes up a different but extremely common view about the intrinsic value of nature. Butterfly, as she prefers to be called, perched herself on the top of a 1,500-year-old redwood tree from December 10, 1997, until December 18, 1999. Far from a tree hugger's caprice, she did so to prevent loggers with the Pacific Lumber Company from cutting down the tree, which she personalized with the name Luna. While aloft, Butterfly conducted numerous press conferences by satellite phone. Almost always, she would emphasize the circle of life and the intrinsic value of the tree that she sought desperately to protect.

Butterfly, of course, is not alone. There are many people in the environmental movement who seek to establish the intrinsic value of nature by appeal to ecology, by appeal to interconnectedness. By extension, many arguments have been made to support these views, some of which are considerably woolier than others. Unlike the *individualist* arguments for intrinsic value that I mentioned in the last section, these *holistic* approaches emphasize *inter*dependence over independence. In this respect, then, holistic approaches are somewhat intermediate or hybrid positions: they seek to identify value as intrinsic to individual bits of nature yet at the same time show how this value is important to the whole, and thus, in a way, extrinsic.

One of the most common and earliest articulations of the holistic interconnectedness argument can be found in Aldo Leopold's important work *A Sand County Almanac*. Leopold excoriates those who think that they can control nature, who think that they can run nature like they might run a machine, who turn a blind eye to the land. Trained initially as a land manager, Leopold proposes that land management would work better if we would accept the ups and downs of the universe, if we would stop thinking about nature mechanistically. He proposes a different ethical standard, a *land ethic*. "A thing is right," he says, "when it tends to preserve the integrity, stability, and beauty of the biotic community. It is wrong when it tends otherwise."[15]

Leopold is not the only one to find value in all of nature. Deep ecologists such as Arne Naess, George Sessions, and Bill Devall have sought to help us understand how intertwined our lives are with nature. According to their reasoning, the mere fact that we are inextricably intertwined with nature should proscribe actions that involve harming nature. We are, they say, all but knots in an intricate web of life, and the self—our core being—is conditioned and supported by a complex network of interrelations. In effect, humans are more connected with nature than disconnected from nature, and insofar as we have interests in living and flourishing, so too ought we to view the rest of living nature as inextricably bound *to* us. We should adopt an expanded view of ourselves. If we value our own lives, the thinking goes, we should value all other living things that make up the various strands of this web—because, essentially, *nature is us*. Harming nature, in other words, in some way harms us.

More than most positions, this one seems to resonate strongly with my undergraduate students. Out of the hands of philosophers, deep ecology quickly morphs into a moderately spiritualized remnant of its theoretical articulation. Certainly, Butterfly's stance exhibits a tendency to amplify our involvement in nature. The blue hominids on Pandora, the fantasy planet in the movie *Avatar*, with magical USB cables in their ponytails, is only the most recent popular-culture instantiation of this attitude toward nature. Partly because of this spiritual dimension, the entire corpus of deep ecology was dismissed as "eco-la-la" by social ecologist Murray Bookchin.[16]

Talk about our interconnectedness needn't be overly mystical or hysterical, however. It can be as pedestrian as the common assumption that nature provides the backdrop and the furniture of our lives. Rachel Carson begins her opus, *Silent Spring*, by asking us to imagine an American town in which the birds do not sing. They have been poisoned to death by the overuse and widespread application of untested pesticides. She wants us to see what a shame it would be to lose this aspect of nature. The thought, presumably, is not only that the world would be deficient without these beautiful songbirds but also that we would miss the birds, that our neighborhoods and lives would be transformed in a way that we ought not to countenance.

The climate community sometimes invokes the interconnectedness theme too, proposing that nature's value rests in how tied it is to our interests. Sully the atmosphere and out goes the climate, they say. When climate shifts, agricultural patterns and migratory patterns shift. As animals and plants begin to respond to shifts in climate, the delicate strands in our finely woven web start to fray, until at last we are left with little but a pile of string where nature used to be. It's not quite the spiritualized account you sometimes encounter with the deep ecologists, but it's the same basic idea.

Almost all of these views, to greater or lesser degree, are backed up by appeals to science, ecology, and an attempt to weave morality out of these otherwise physical dependencies between entities. But clearly, ecology is not enough to make the case that we should change our behavior toward nature. Just because there are some things with which we are intimately interconnected doesn't at all suggest that we should simply accept them if they appear to be bad for us. We cut out cancerous tumors, for instance, even though they are deeply intertwined with our bodily systems.

Furthermore, simply noting the interconnectedness and proposing that somehow this isolates the intrinsic value of nature doesn't provide us with many clear answers about what to do, which is ultimately the purpose of ethics. I am connected to my neighbors, but it's not clear from the mere fact of this relationship how I should treat them. Thus, there is an action-guiding problem here.

To see how vague and ambiguous it all is, consider practically how our observations about interconnectedness have actually played out. It seems that the more we know about the interconnectedness of germs and pathogens, the more concerned we are about what we ingest and how we ingest it. This is, of course, good—up to a point. But almost every public building now is lined with dispensers of hand sanitizer. Public health demographers warn that overuse of antibiotic soap may allow for the proliferation, not the annihilation, of more virulent pathogens. Antibiotics present the same problem. We sterilize like crazy and give antibiotics to our children at the very whisper of a dread disease. These early epidemiological directives are, in part, a direct result of our continued acknowledgment that we are intertwined with nature.

Maybe you think this is just a problem with our information, that we would act differently if we knew more about the path of disease or antibiotic resistance. But I think that's probably wrong too. We still face free-rider problems: some people will benefit from the herd immunity of others even though they themselves do not get vaccinated. And we also face tragedies of the commons: if I am sick, I need to use antibiotics; I am not about to sacrifice my own health for the common good. Other people oughtn't to, either. Over time there just will be more virulent pathogens.

So I have concerns about the extent to which these intermediate holistic approaches to valuing nature can be action guiding. Which brings us to the other possible class of arguments: the extrinsic.

BEES AND THE BOTTOM LINE

During one particularly contentious forest campaign, I and at least 20 fellow activists planned to blockade a logging road, aiming to stop the wholesale clear-cutting of a privately owned forest in the

Southwest. In the middle of the night, we set up a tripod and two cement barrels in the road, locking our arms and necks to the barrels. On the morning of the action, as the warm sun rose over the sage flats and the logging trucks barreled toward us, I was gripped by a sudden sense of the real-world implications of my environmental stance. This was no theoretical battle between the forces of good and evil. This was real.

At first the caravan of trucks revved their engines and sped toward us. Approximately five feet short of where we stood, the lead truck stopped abruptly, its pneumatic brakes releasing with a pop and a slow hiss. The door opened. Out jumped a man—a logger, from the looks of him—quite a bit bigger than me, but also, clearly, a human being. He bounced up to me, anger and frustration pulsing in his neck. There was no mincing of words. "Move that fucking thing out of the road!" he yelled, finger wagging in my face. I gulped. I apologized. "We're not moving," I explained. "We're here to stop you from logging this forest." Other men hopped out of their trucks. A pickup roared toward us from a side road. The taste of dust could have scratched enamel off teeth. "What's it gonna be," he growled, "my job, or your damned trees?" Dutifully, I spun my yarn: *our* trees.

Many environmentalists instinctively accept that "jobs or trees" is the proper question for the environmental debate, and then they respond by loudly proclaiming how business-friendly green can be. "We can work together," they continue. "The destruction of the natural ecosystem is bad for you. It's bad for all of us. Just a few minor adjustments, and we can build a green economy." We've heard it a million times.

The familiar idea, of course, is that nature offers us ecosystem services, much like the burger flipper at the local fast-food restaurant provides us with burger-flipping services. For years we have relied on these ecosystem services to help us survive. All the while they've remained in the background, benefitting us tremendously, and we haven't paid particularly close attention to how reliant we are on them.

Bees are very good pollinators of flowers and fruit trees, the reasoning goes; far better than we could ever be. Pollination of this sort is necessary for most of our food. We would be fools to try to outdo nature in this regard. Pollination is a classic case of ecosystem services, but the notion of ecosystem services extends well beyond this. Another classic case involves the Catskill Watershed, which allegedly offers untold economic benefits to the citizens of the state of New York by filtering and purifying water.[17] The point has been extrapolated to apply to a huge range of other important ecosystem services. In 1997 economist Robert Costanza and his team estimated that ecosystem services account for $16–$54 trillion per year, stating that such an estimate was probably on the conservative side.[18] Economists like Gretchen Daily and Katherine Ellison write eloquently about the extent to which conservation can be profitable.[19]

This line of reasoning, however, is deeply problematic, or at least contingent upon a bunch of other factors.[20] It certainly cannot serve as a catchall for the overarching value of nature. It is not the case, for instance, that nature provides better shelter for us than we can provide for ourselves. If we rely on nature for our shelter, we'll likely be living in caves and hollowed-out trees.

More problematically, arguing the extrinsic economic benefits is actually very much at odds with conventional environmental wisdom. Some recent work on the economics of climate change suggests that economic and ecological interests may not line up at all in the face of climate change. If you talk to a climate scientist, the problem with climatic change is that it could lead to the collapse of ecological systems. Glaciers will disappear, coastlines will flood, crops will go sour, agricultural practices of a given region will have to be revised, migratory patterns of birds and insects will shift northward, and species of animals will become extinct. Biologically and geologically, it does appear to be devastating.

But talk to environmental economists and you'll get an extremely different answer. They see the world, and the harm brought about through the changing climate, not in terms of the disappearance of geological formations, shifting climatic patterns, or changes in the flora and fauna but in terms of resources, in terms of goods and substitutable goods. In fact, even the most progressive and concerned environmental economists don't anticipate the same sort of catastrophe that the climate scientists are predicting. They're talking about a different kind of catastrophe.

The Stern Review report on the economics of climate change—a 700-page behemoth document released in late 2006 by the British government and written chiefly by renowned economist Nicholas Stern—projects a 5–20 percent decline in gross domestic product (GDP) "now and forever," if all the costs and risks associated with climate change are factored in.[21] Certainly, from the standpoint of growth, that's a really bad thing, and we ought to take steps to protect ourselves from it. It is also catastrophic—a great depression to beat all great depressions. But notice that this is only really an anticipated slowdown in growth. The impact is on the economy, not on the earth. The economy will still grow, and people will still be relatively well off, they just won't be *as* well off as they currently are. This is a far cry from the Jared Diamond-style ecological collapse that has so many environmentalists and climate scientists concerned.[22] Wealth will still be created, but it'll just be created at a slower rate. That's not nearly as terrifying.

To complicate matters, talk to different economists of an arguably more conservative nature and you're likely to get even more optimistic projections. Richard Tol and William Nordhaus, both in different reports, suggest a much smaller decline in the economy and GDP due to climate change.[23] In part, this is due to their different discount rates, which is the rate at which we make sense of future value. A very aggressive discount rate operates on the assumption that a dollar today is worth quite a bit more than a dollar a year from now, whereas a far more conservative discount rate operates on the assumption that the value of a dollar a year from now is closer to its value today. Tol and Nordhaus use this very aggressive, optimistic discount rate, valuing the present much higher than the future. The *Stern Review* uses the extremely conservative, pessimistic discount rate, valuing the future more on parity with the present.

What you must notice, however, is that both of these two views about the impacts of climate change—the one that says that we are facing dramatic ecological collapse and the other that says that it is our economy that will collapse—emphasize harm and/or damage to something of value. What they don't do is clearly link those values together. A conservation biologist may be very upset about the climate-induced loss of the American pika or the giant panda. An economist, in contrast, may be upset only insofar as there is diminished wealth in the economy. The important difference is not in argumentative strategy; it's in what is being harmed and/or valued. This really matters, actually, because if your concern is primarily with wildlife, there may be concrete steps you can take to protect that wildlife; but if your concern is primarily with reduced economic productivity, you will be drawn toward a very different set of concrete steps, which may pay very little attention to wildlife.

GREEN IS NO GOOD

Just as the environmental movement has shifted its sights over the past decade, so too have I.

My life and my perspective have changed substantially since my early days as an environmentalist. I'm no longer a graduate student, for instance, so I no longer feel as powerless as I once did. I've also spent a fair bit more time living in large cities and speaking with social justice activists.

Seven years ago I was in Copenhagen for the UN Framework Convention on Climate Change meetings, this time on the other side of the protest barriers and police blockades. Like many around me, I wore a suit and dangled credentials from my neck. I watched cautiously as protesters launched guerrilla theater projects and shouted expletives at the amorphous specter of global capitalism. Indeed, as the meeting on international climate policy threatened to unravel, the international discourse centered on the economic challenges that a binding cap-and-trade regime—a central policy mechanism under discussion—would pose to US jobs.

The discussion continues apace. Neither was the climate policy problem resolved in Cancún a year later, though a slightly more congenial outcome emerged from those talks. The Cancún Agreements of 2010 brought 193 nations together to help developing countries cut their carbon emissions. Throughout the negotiations, the rhetoric was heated. Detractors on the right claimed that the cap-and-trade regime will cost billions in lost opportunities.

Detractors on the left claimed that cap-and-trade is a giveaway to polluting interests, that it only kicks the can down the road, that it makes no serious attempt to move us in the direction of a sustainable future.

The point of this long-winded disquisition is not only to isolate the various arguments that animate the environmental movement, but also to call attention to the way in which the precious vase view—whether supported by intrinsic or extrinsic value arguments—underlies an incredible range of environmental thinking. Here we have two core approaches to environmentalism, each represented by a broad spectrum of substantive arguments, both of which emphasize or speak positively about some valuable aspect of nature. In almost all of these cases nature is said to be good, to have unrecognized or disregarded value. In a nutshell, this is motivated by the logic of the precious vase view, which I presented above.

Given the nature of capital-driven, postindustrial democracies, there is a sense in which assigning a value to nature flattens the normative discussion to concerns primarily about pricing and ranking. What if, in the face of the current global economic crisis, we decide that nature isn't, in fact, worth preserving? What if we decide that to avoid an extraordinarily painful recession, it is an undesirable but necessary alternative that we leave our children a slightly degraded environment? Suppose we frack or drill our way to a more robust economy. We appear to have few qualms about borrowing trillions of dollars from our children's financial future, so why not instead exploit our natural resources as a method of reducing the economic debt that we will leave them?

Intense resource extraction has always been an easy fallback, and once again it could provide trillions of dollars to the national treasury. Whole forests could be razed, whole fisheries could be destroyed, whole habitats disrupted, simply to keep us from suffocating under trillions of dollars of debt. Nature's bounty could provide a reasonably painless bailout, the effects of which would not be felt for several generations. For at least the past 10 years the standard environmental line has been that we should live sustainably because we cannot leave our children, or our children's children, with global environmental devastation. But we are reminded daily that we may face similar, perhaps worse, devastation if we allow our economy to unravel. The financial stakes in this game are enormous. The environmental stakes are equally large.

I suspect that this pronounced love of nature, this precious vase view, has been the environmentalist's Achilles' heel more times than most of us care to admit. Certainly, being called a tree hugger is generally not positive. It's a put-down, aimed to sissify and infantilize the environmentalist. It is also ripe for dismantling. Countless written refutations of environmentalists point primarily to the horrors of malaria, plague, the necessity of pesticides, and nuclear energy as evidence that treehuggery is tantamount to wingnuttery.

All this to say: the environmental debate is frequently framed as a battle between frosty skeptics and squishy environmentalists, between nature haters and nature lovers. That's been the predominant trope for decades. If this trope is not upended once and for all, before the new green becomes the old black, we very well could find that the sheen has worn off our shiny new locomotive. Put differently, we're poised now to make the same mistakes that have had environmentalism languishing from the 1960s until the zeds. It is time to carve out an entirely different approach to being green—an approach that isn't reliant on the familiar rhetoric of the early environmental movement. To do that, however, we're going to have to acknowledge a deep secret.

True, the earth gives much to us, including sustenance, care, and shelter. It gives us mountain prairies and aspen stands, vast oceans to explore, rolling tundra, incredible light shows, diverse and exciting species, and much more. There are a million things in nature to love. I like my fresh tomatoes and my hiking trails just as much as the next guy. But if there's one prevailing thread that runs through this book, it is that nature also gives us fair indication that she doesn't so much care for us.

For all of the benefits that nature confers on us, for all of the pleasures and joys and beauty of nature, for all of the purple mountains, amber waves, and orange skies, nature has a dark side.

3

Rustling in the Bushes

In which the author (1) reminds the reader of the horrors of nature, (2) explains that the dark side of nature has always terrified us but that scientific discovery has helped to shed light on the "out there," (3) proposes that where science has offered clarity, technology has helped to control and tame nature, (4) suggests that nowadays science and technology thoroughly condition our experiences of nature, and that this is neither necessarily good nor necessarily bad, (5) reiterates that despite our conditioning technologies, nature and its hazards are inescapable, and (6) proposes that we can still gain a grip on how to think about nature if we ignore the question of nature's value or, more dramatically, even if we assume that nature is bad.

HORROR

When the Ives family set out on a Father's Day camping trip in the American Fork Canyon of Utah, just fifteen minutes from their home, they could little have anticipated what lay in store.

Their camping experience began normally enough. The family of four pitched their tent, ate their dinner, told stories by the campfire, and went to bed. At around eleven o'clock they woke to hear their eleven-year-old son, Samuel, screaming hysterically: "Something's dragging me!" he shouted. Those were the last words his family heard. They never saw what happened. Police later found Samuel's body approximately 400 yards from the campsite. A black bear had torn through the wall of his tent, dragged him out by his head, and mauled him to death.

Since then, as one might expect, the Iveses have been left wondering what they might have done to avoid this tragedy. What had they seen? Had they done anything wrong? Could they have been more careful with their food? Had they foreknowledge of the bear, would they have chosen a different site? Should they have brought the dog? Why didn't they bring a gun?

They filed a lawsuit against the US Forest Service, claiming that they had not been warned of the bear danger, that they would have acted differently had they known that there was an aggressive bear in the area. "Though rare, bear attacks happen," the rangers responded. "The Forest Service cannot be held accountable," the lawyers replied. Scores of well-meaning friends and neighbors have undoubtedly consoled the grieving family by explaining that they couldn't have known what was going to transpire. Still, the Iveses persist. No doubt they have replayed events over in their minds. Different information, they contend, would have resulted in different actions.

It's hard to say whether this is true. Maybe it is, maybe it isn't. Their reaction, however, points to a phenomenon that we experience all the time, particularly after tragedy strikes or mistakes are made, but actually with many of our decisions: we retrace our actions and related events in search of an explanation for what happened.

It is a process not unlike the explanatory soul-searching conducted by Buckland, Mantell, Owen, and Cuvier, who sought answers about what happened to the dinosaurs, and it is similar to the explanatory prognosticating employed by our current crop of climate scientists who want to know what will happen with our climate. In this case, however, Samuel Ives's parents weren't seeking a strictly causal explanation. They were seeking a *modal* or *counterfactual* explanation: they wanted to know what *would have* happened *if* ...

• • •

Most rangers will assure you that bear maulings are relatively uncommon. There are only two or three fatal bear attacks per year, though there are a larger number of attacks in which the victims survive. But it would be a mistake to confuse information about past bear attacks with the wider danger that nature poses to us and our children. Bear attacks cannot be considered independently of other natural hazards. We're talking about nature, after all, which includes basically all things that prey on or pose a threat to us. An alligator is a kind of predator, as is a tiger, a dingo, and a shark. I'm sure it's perfectly reassuring that you won't be eaten by an alligator as you're hiking in the woods, but watch out for mountain lions, bobcats, raccoons, coyotes, snakes, and wolves, and take all their attacks to heart. While you're at it, beware of ticks, wasps, bees, rocks, cliffs, lightning, falling branches, and the occasional pit of quicksand.

One study estimates that there are approximately 4.5 million dog bites per year in the United States alone, requiring millions of dollars in medical treatments.[1] The number climbs into the tens of millions when one expands the inquiry to include dog attacks worldwide; and when one factors in attacks from other animals, including the bites and stings from venomous insects, the number grows unfathomably large.[2]

In a related longitudinal study examining data from 1979 until 1990, researchers evaluated fatal animal and insect attacks in the United States. The categories of attack read like a laundry list of parental nightmares: venomous snakes and lizards; venomous spiders; scorpions; hornets, wasps, and bees; centipede and venomous millipedes; other venomous arthropods; venomous marine animals and plants; dogs; rats; nonvenomous snakes and lizards; cats; moray eels; sharks; rodents other than rats; monkeys, and so on. The researchers also examined cases of more gruesome attacks, including being butted by an animal, fallen on by a horse, gored by an animal, stuck by the quills of a porcupine, pecked by a bird, run over by an animal, or stepped on by an animal.

Though most animal attacks are nonfatal, many studies suggest that animal attacks account for tens of thousands of deaths per year worldwide. The 11-year period of the above study resulted in only 1,882 fatalities: 66 from snakebites; 49 from spiders; 527 from hornets, bees, and wasps; 186 from dogs; 796 from unspecified animals.[3] Mountain lions? From 1970 to 1990 there were 31 nonfatal and 5 fatal attacks on humans.[4] Alligators? The Florida

Department of Fish and Wildlife claims that since 1948 there have been approximately 360 attacks on humans, only 25 of which have been fatal.[5] Dingoes? Well, one attack, if you agree with researchers that the dingoes at Ayers rock ate poor little Azaria Chamberlain more than 30 years ago.[6] If we include in our count the number of mosquito bites causing malaria deaths, the number jumps another 1.5 to 2.7 million, depending on the year. That's a really big number, and one that we'll return to in the coda of this book.

To be honest, I find these numbers both terrifying and reassuring. They are terrifying because I can scarcely imagine what it must be like to hike along a trail or play in my backyard, lost in my thoughts, only to be suddenly yanked back to reality by a creature much larger than I. Like a fish swimming naively up to a tasty worm only to be greeted by a hook and a painful tug, one moment you're minding your own business, the next moment, you're dead.

But the numbers are also reassuring, since the odds of any particular kind of attack are relatively low. Environmentalists are, in a way, correct to suggest that animal attacks are relatively rare. Fact is, if you're playing in your backyard or sitting down to stick a chicken wing in your mouth, it's unlikely that Mother Nature's fishermen will violently yank you out of your chair. But this too is deceptive. Animal attacks are rare not because nature rarely attacks but mostly because we do our damnedest to control nature and keep her at bay.

No question, many of nature's wild creatures are majestic brutes. They are beautiful. They are magnificent. And maybe they are even wonderful or great. But, as Edward Abbey once said, if people continue trespassing in grizzly territory, there can be little doubt that, from time to time, the grizzlies will harvest a few trespassers.[7]

Abbey aside, it seems to me to be a contortion of the highest order to suggest somehow that the expression of a bear's nature in the form of the brutal mauling and killing of Samuel Ives is in any respect beautiful, wonderful, or magnificent. It is as horrible an event as I can imagine. Every time I am in the woods camping, alone or with my son, I think of this event. I think of Samuel and

I think of my son; I think of Samuel's family and I think of my family. This single anecdote has lost me many hours of precious sleep.

• • •

Most of us experience at least one tragedy over the course of our lives, and most of us have felt the danger that nature poses. It is likely that our tragedies are not as intimately tied to nature's more charismatic representatives like bears, but in the end, if not a car or a gunshot or a smoking habit, nature, red in tooth and claw, does most of us in.

The dangers posed by nature compose the central problematic of this book: If nature can be both violent and beautiful, what about nature should lead us to revere it? What about nature should lead us to protect it? More important, if we cannot find value in nature, or at least acknowledge that raw nature is overflowing with as much disvalue as value, then where else can we look to unearth a better understanding of our responsibilities *to* nature? In the face of stories like the Ives boy—and there are plenty of them—one might be inclined to throw up one's hands, take a defensive stance against nature, and do whatever possible to combat the perils of nature in order to avoid the negative outcome of bear attacks (or tornadoes, or tidal waves, or disease, or anything).

In the previous chapter I touched on two approaches to nature's value: distinguishing between intrinsic and extrinsic approaches. There I was concerned to isolate several prominent strains of thinking about the *source* of nature's value or, as philosophers say, the source of the good. Here, however, I want to discuss how reliance on the idea that nature is valuable, whatever the source, falls flat on its face. While there are many ways to approach this question, I'll focus on just one: the extrinsic view, the idea that nature is good *for us*.

Before I proceed, however, I want to be clear that it's not that I don't think we should promote or protect nature—I do think that we should promote and protect nature—it's that I don't think we should promote or protect nature *because it is good*. In fact, I think we should protect nature *even if it is bad* (but not *because of* its badness). I want to show in this chapter that nature can be quite bad. What I actually think, and what I will argue here, is that nature is neither good nor bad, it is only ever wild. This, I presume, is an essential piece of the environmentalist's puzzle; a piece that has more or less been missing from the core environmental discourse.

TERRA INCOGNITA

Even though the statistical incidence of bear maulings and alligator ingestions is relatively well researched, nature still has a few secrets up her sleeve. When the startled populace of Puerto Rico woke one morning in 1987 to read in both major newspapers, *Il Vocero* and *El Nuevo Dia*, that a mysterious creature had been eating local domesticated animals and leaving their carcasses behind, the mystery was simultaneously new and familiar. Chupacabra (literally, the "goat sucker") was said to sneak into fields at night, chow down on livestock, and retire into the shadows to digest its crimson victuals. Rumor spread quickly, and within a few years the chupacabra was sighted in locations around the world, raising the specter that it was not one lone creature, like Bigfoot, but a species of creatures.

Unlike most species, however, regional variations of the chupacabra are wide-ranging. In parts of Puerto Rico the creature is said to be a half lizard, half kangaroo hybrid with red eyes and sharp fangs that hops over fences and into barns under cover of darkness. In Texas it has been described as a hairless wolflike creature that skulks under bushes and lives in the shadows. Reports from Spain describe it as lizard-like, with quills on its spine and a long, snake-like, forked tongue.[8] In all manifestations, however, the chupacabra is an enigma, flummoxing farmers and villagers and thereby falling into a class of mysterious and mythological creatures better described by their vexingness than by their taxonomical attributes: *cryptids*.

Several other cryptids are equally well-known: the Loch Ness Monster, Bessie, Sasquatch, the Abominable Snowman, and the Montauk Monster. These are the superstars of the shadows, those who emerge to spook the neurotic, titillate the curious, and terrify the young. But there are many other lesser-known cryptids. There's Ahool, the flying giant bat, rumored to live in the rainforests of Java; the Ebu Gogo in Indonesia, a tribe of three-foot-tall hominids said to have long arms and potbellies; and the Beast of Busco, a mutant snapping turtle that allegedly lives in the swamps of, of all places, Indiana.[9]

They get better: Chicken Man, the Fouke Monster, Orange Eyes, the Bear Lake Monster, Ogopogo, the Nagumwasuck, the Earwig, the fur-bearing trout, the sand squink, the snipe, the squonk, the Terichik, the Wi-lu-gho-yuk, and many, many more. All, no doubt, have kept many young campers awake at night wondering what that scratching, breathing, huffing outside their tent is. Hearts pounding, eyes shifting wildly, millions upon millions of kids have screamed from their sleeping bags, "Get. Me. Out!"

Amusing though tales of cryptids may be, they are no contemporary phenomenon. Hobgoblins have been a concern of all who have ever stared into the darkness and flinched at an unfamiliar or unexpected sound. For most of the history of civilization, nature has been an elusive foe, lurking treacherously around every corner. It has hidden in the bushes, waiting for our guard to come down, anxious to strike when we least expect it. The dangers of nature have taken many shapes, appearing and reappearing as animals, natural events, toxins, pathogens, and even bad weather, but our sentiments could be summarized no better than in the various bogeymen that have peppered our legends and literary lives.

The Anglo-Saxon King Beowulf battled the monster Grendel, said to be a descendant of Cain. Odysseus had to sail between the choppy waters of Scylla, a three-headed sea monster, and Charybdis, a creature so bloated that it existed almost entirely as a disembodied mouth. In the *Aeneid*, the three-headed dog Cerberus watches over the gates to Hades, waving its serpentine tail, distractible only by a wine-soaked sop of bread. These mythological cryptids are frequently serpents, or lions, or dogs, and sometimes they have multiple heads, but they almost always have some human features. Unsettlingly, they are almost always intelligent.

Monsters have been the subject of great scholarly interest as well. The great grandpappy of philosophy, Aristotle, developed a complete theory of monsters.[10] In a work compiling the history of animals, an early precursor to the bestiaries of the Middle Ages, Aristotle describes monsters as a "mistake of purpose" in nature.[11] Sir Francis Bacon, arguably the chief mason in laying the foundations of the scientific method, writes that nature exists in three states, at least one of which occurs when "she is forced out of her proper state by the perverseness and insubordination of matter and the violence of impediments."[12] What is that state? Yep, you guessed it: monsters.

Thomas Hobbes's magnum opus is in part a long-winded homage to a metaphorical sea monster. Though he never references the spooky water dweller directly and instead details the conditions under which men behave as animals—conditions like equality, scarcity, and self-preservation—for Hobbes, life in the state of nature was solitary, poor, nasty, brutish, and short.[13] Even the great theorist of noble savages himself, Jean-Jacques Rousseau, who otherwise thought nature quite good—indeed, says that "everything degenerates in the hands of man"—felt that some denaturing of man was required for civilization.[14]

Fortunately for all of us, the scientific method has helped to defang some of nature's gnarliest beasts. William Buckland's simple act of giving Anning's dragon a name may have been, in at least the case of the dinosaurs, enough to do the trick. Many other rumors and myths have undoubtedly been laid to rest by careful scientific investigation. Our field scientists have helped to take the mystery out of most things that go bump in the night. The demystification of nature has helped us get past our primal fears. We can now safely open our closet doors, peek under our beds, and see what has been there all along: piles of dusty clothing.

Despite all that science has contributed to take the bite out of nature's beasts, nature's mysteries continue to manifest, even in our more contemporary literary outlets—basically, movies and television. They've morphed a bit, but they're roughly the same: enigmatic creatures inhabiting the space of the inexplicable and the unknown. The monsters of the 20th and 21st centuries typically originate from the nether regions of the cosmos, or sometimes as the result of scientific experiments gone awry. They are aliens, predators, mutants, hybrids, diseases, and other features of nature that we do not understand and cannot control. And, as ever, they are almost always intelligent and empowered agents. Even as we have come to understand more about nature, our foes continue to outwit and elude our scientists.

All of which is to say, in short, that even our most advanced scientific knowledge is not enough to take the creepy and the scary out of nature; that no matter how much we believe that knowing and understanding our world will release us from our night terrors, nature will continue to haunt us. Simply knowing more has only ever been half the battle. Nature, now as then, is something to shut out, to rally against, and to fight.

RAINBOW OF FLAVORS

On October 21, 1967, *Washington Star* photographer Bernie Boston snapped a young peace activist named George Harris as he placed a carnation in the business-end of a military rifle. This grainy black-and-white image was so powerful that it all at once challenged soldiers to embrace peace, called attention to the curative and restorative powers of nature, and inspired generations afterward to regard the "flower children" as petulant infants, naive to world affairs. From the peace movement—with its emphasis on flowers and herbs, natural living, and free love—the environmental movement would be born, inspired by many of the same ideas and motivated by many of the same players. What predated the photograph, of course, was the Vietnam War, and what predated the Vietnam War, or at least fell smack-dab in the middle of it, was the publication of Rachel Carson's work on birds and pesticides.

This may seem a somewhat tenuous relationship, but the two events can't well be understood independently of one another. Nature was an enormous problem for American troops in Vietnam, and Vietnam was an enormous rallying point for the fledgling environmental movement. If you think about it, it makes perfect sense.

The jungles of Vietnam offered notoriously difficult terrain and inclement weather that made the days and nightmares of war even more hellacious. The long rainy season, from May until September, kept the nation at a humidity level of about 84 percent throughout the year, which facilitated dense growth. Thick foliage camouflaged the spider holes, tunnels, and weapons of the Vietcong. The monsoon rains caused disease, mildew, and the deterioration of clothing, tents, and boots. The eight-foot-tall elephant grass tore neat slices in bare arms and legs. The jungles of Vietnam were no friend to American soldiers.

The terrain was entirely unknown to us, which was our great disadvantage. To combat the problem that nature was posing, we turned to the nearest source of knowledge—the sciences—to aid us in developing technologies that would give us the upper hand. The US government instituted an aggressive program of defoliation. For 10 years, from 1961 until 1971, the military dropped caustic herbicides on the forests and jungles outside villages suspected of harboring Vietcong. This program of herbicidal warfare included use of the oft-vilified Agent Orange, a chemical cocktail including dioxin, one of the more carcinogenic substances known to man.

Of course, Agent Orange wasn't the only colorful chemical in the US arsenal. Its intrepid efforts were aided by others. The crew of chemical superheroes in US tankers came in a rainbow of toxicities: Agent Purple, Agent Pink, Agents Blue, White, and Green. It is unclear how Agent Orange was crowned the potentate of this illustrious crowd, but these bit players in the Justice League of Plant Killers composed the cheerful band of Rainbow Herbicides.[15]

To some extent, this was the first case of greenwashing: an attempt to downplay the environmental hazards of a given product or activity by giving it the appearance of environmental friendliness. Before invoking the innocence of rainbows, the program was known by the far more menacing moniker Operation Hades. As it progressed, however, the operation's name was buffed up for political reasons, transforming first into the considerably less caustic Operations Ranch Hand and Trail Dust, only to settle, eventually, on rainbows. The brutal naming irony invoked American lore of the western cowboy, except that the dust that these ranch hands kicked up contained a lethal cloud of dioxin.

In total, the US military dumped more than 72 million liters of the Rainbow Herbicides over the jungles, the enemy, and innocent bystanders.[16] War theaters being what they are, the herbicides were sprayed not only on bothersome plants and animals but also on our very own soldiers. Not long thereafter, soldiers began to experience bouts of tingling nerves and motor weakness. The long-term effects of Agent Orange, largely unknown at the time, are really horrible. They include acute peripheral neuropathy, amyloidosis, chloracne, chronic lymphocytic leukemia, diabetes mellitus, Hodgkin's lymphoma, non-Hodgkin's lymphoma, ischemic heart disease, multiple myeloma, Parkinson's disease, porphyria, prostate cancer, respiratory cancer, and soft-tissue sarcoma.[17] All told, more than 3 million people, both soldiers and civilians, were exposed to the Rainbow Herbicides, the effects of which still linger today.

Agent Orange and the war against nature soon became a rallying point for the fledgling environmental movement. Carson's *Silent Spring* called attention to the indiscriminate use of herbicides and pesticides on the domestic agricultural front, but she artfully deployed the terminology of war to build her case. She wrote of "catastrophe," the "decimation of bird populations," and "rivers of death." In his afterword to *Silent Spring,* Edward O. Wilson so much as referred to the spraying of the insecticides dieldrin and heptachlor as the "Vietnam of Entymology."[18] Though *Silent Spring* is far more focused on DDT than on Agent Orange, there can be little doubt that this monumental work resonated strongly with the American public precisely because the US military's war with nature was simultaneously going strong on the other side of the ocean.

To no one's surprise, *Silent Spring* generated an immediate outcry from vested interests.[19] Chemical companies like Monsanto, Velsicol, and American Cyanamid began a heated attack on Carson's position, filing lawsuits and torching a parade of straw men by suggesting that Carson's view was deleterious to public health.[20] To this day, industrial criticism continues. By laying the groundwork for the banning of DDT, they argued, Carson singlehandedly permitted more malaria than any environmentalist.[21] Though this view is demonstrably false, the charge is repeated again and again.

Of course, the United States was hardly the first to institute a program of herbicidal warfare. Such warfare has been practiced for centuries, generally with much less technical sophistication. Before the unicorns and sparkles suggested by the Rainbow Herbicides, herbicidal warfare employed one of the most destructive and primitive forces known to humankind—fire. Scorched-earth defoliation campaigns were practiced by almost all successful armies: the Persians, Greeks, Romans, Gauls, Vikings, English, French, Germans, Russians would leave carbon black coals where villages, farms, and forests had once stood.

The great irony in scorched-earth warfare is that in all cases soldiers have been reliant on nature even while struggling against it. To the American soldier, the elephant grass of Vietnam was a menace, masking the true dangers that lurked beneath. To the soldiers of the Vietcong, however, the elephant grass provided life-preserving cover. Nature's disservices for one party were very much nature's services for the other party. Nature was, in effect, a mere medium in which the war was to take place, indifferent to the outcome of the war or the justifications of politicians. It generally provides its services and benefits to those who are familiar with it. American soldiers during the Revolutionary War were the beneficiaries of the once lush east coast forests, where British soldiers were its victims. Much of what it means for nature to provide us with an ecosystem service or disservice is contingent on what we want it to do for us.

As we have grown more knowledgeable about the plants and critters outside our towns, we have also grown more adept at putting nature at our service, both by using it to our advantage and by keeping it at bay. For centuries we were limited to horticultural grafting and selective breeding to improve our agricultural and livestock diversity and quality; now we're not only in the business of genetically modifying our seeds to resist weeds, we are also in the process of genetically modifying our animals to produce more milk or meat. Some technologists are even working on gene therapies for humans, perhaps to cure ailments but also to improve upon what we have been given by nature. It continues apace. The corporate titan that now provides us with weed killer and herbicides to grow crops and feed our children, Monsanto, is the same company that once upon a time, back in the day, manufactured Agent Orange for the US government and fought Carson's crusade against DDT. Monsanto and other companies are also now busily engaged in the production of genetically modified organisms—crops, in many cases, but animals, even chimeras, in other cases—effectively utilizing recent scientific discoveries and the accumulated knowledge of the biological sciences to create man-made versions of the very same cryptids that more than once have scared the hell out of children.[22]

Filtering nature through the lens of war, it all begins to make much more sense how environmentalists first came to the conclusion that the core environmental problem was an attitudinal one—that we simply need to modify our attitudes toward nature; that we must open the closet doors and cast light on the various supernatural bogeymen known to lurk in the shadows. Whereas most of this history of civilization has been hell-bent on the eradication of the perils of nature, and has even deployed policies like scorched-earth warfare, it would appear that the appropriate response to these problems is to show how important nature is, to illustrate how valuable the birds and the bees are. As I've been arguing, however, this view is mistaken.

Science has demystified the earth, taught us more about our environment. Technology has empowered us to harness and control it. We've moved, in this respect, from an attitude of loathing toward nature to an attitude about how best to use it for our purposes. As we've learned more, nature has lost its mysterious and supernatural edge. Tales of the chupacabra still pop up now and again, but there's usually a better explanation that can quell our fears, just as there was with Mary Anning's stunning megalosaurus find. Many researchers now believe that the various sightings of the evasive chupacabra can be explained as wild dogs or coyotes suffering from a severe form of mange.[23] As we've learned more about nature, we've grabbed the reins of nature, steering it to do our bidding. In so doing, we've further blurred the boundary between the world out there and the world we inhabit.

ELEVATORS AND AIR CONDITIONERS

My great-grandfather Maynard had only 9 fingers. He lost his 10th in a table-saw accident. Most of his fingers —all nine, that is—were standard-issue ticklers: long and bony. The 10th was a hideous stub. Like a few other grandfathers, he could play that creepy grown-up trick of pulling off his ring finger in an incredibly convincing way, leaving my cousins and me open-mouthed and wide-eyed.

My relationship with my great-grandfather seemed to me mostly mediated through his deformity. I vividly remember watching him gesture this way and that, while I would fixate on his stub as he'd move his arm to make a point. I recall recurring childhood nightmares in which he would suddenly have all 10 fingers, or in which I was forced to distinguish between the true Maynard and a 10-fingered impostor. My private knowledge of his not-so-secret deficiency would help me prevail in choosing the right man.

In some respects he was a paladin. He was a skilled marksman, an accomplished hunter, and a tireless fisherman. Parched from years in the sun, his ruddy skin had the feel and look of the crunchy brownish paper that protects the inner layers of an onion. His arms would fold and crinkle as he'd wave his right hand in my face, puffing on his pipe, to display the mangled stump of his ring finger. He'd seen a lot, done a lot, hunted and camped a lot; as a result he came equipped with a wealth of wisdom to impart to his great-grandchildren, which he did with aplomb. Above all, my great-grandfather violated what I am sure is an ironclad rule of grandparent superheroes: he once let me in on the secret to his outdoorsy ambidexterity.

"Any old jackass can go out in the wilderness and be miserable," he told me. "It takes a little planning and ingenuity to have a good time."

My early experiences with camping were thus exercises in making our temporary sleeping quarters into a miniature replica of our home. Invariably we would pack a 10-person tent, even though there were just a few of us, a camping stove, several flashlights, lots of bug spray, two coolers, inflatable air mattresses, tap water in an enormous plastic bladder, an electricity generator, a fan, and a television set (!)—not quite the naturist's ideal, but a fine time in the outdoors nevertheless.

Now that I'm older, my camping trips are less elaborate. I've scrapped the television and many of the bulkier items, though inevitably I bring my digital camera, a Nalgene bottle, a knife, a hatchet, a lantern, a stove, some really puffy sleeping pads, and so on. I use a tent, of course. I wear bug spray. I conduct tick checks. I hang my food to keep it from bears, and I still get scared at night. I also have a special place in my garage for all my equipment, which involves quite a few more goodies than will actually fit in my car, including a second gas stove, a smaller tent, a thick sleeping bag, several other thin mattresses, and pots and pans so dirty that simply to touch them is to run the risk that clothing two rooms away will mysteriously be damaged by soot.

I'm not alone. When unwashed forest activists gather on mountains to protest the destruction of nature, almost everyone brings an incredible amount of camping gear: tents, camping mats, lanterns, Gore-Tex gloves and boots, waterproof tarps, plastic water containers, water purifiers, distilled iodine tablets, lanterns, cookstoves, and pots and pans. Almost everyone's enjoyment of nature is ineluctably mediated by modern life. Our experience of nature is conditioned by our technologies, our human interventions. So much of our life is permeated by "air-conditioner" technologies that we can scarcely imagine nature without them. We hardly blink when we pass well-manicured lawns, glimmering paint, perfect sidewalks, unblemished fences, sparkling fields, and waves of grain. The technologies that keep our lives running smoothly are employed in a constant battle against the elements. Some theorists call this new habitat a "second nature," the artificial jungle upon which we are now dependent.

Fact is, our environment is extensively conditioned by our technologies, from our air conditioners to our refrigerators to our cars, and let's not limit it to the gadgets that grace electronics stores. Conditioning technologies include any devices and tools that we use to improve our lives, from abacuses to aqueducts. Simply putting our underwear on in the morning is enough to keep the bugs from biting our bottoms. To go camping in this age amounts to little more than turning down the air conditioners, leaving a precious few of our amenities at home and facing nature with a few fewer, or as few fewer as we'll accept, conditioners as possible.

Of course, it has always been true that technology's core function is to control and steer nature to do our bidding. No doubt our medieval forebears were the beneficiaries of whatever technologies existed in their time. Almost all of what we generally consider to be modern civilization has been built with the idea of taming some aspect of nature. Remember those animal attack statistics I cited earlier? They're low in part because we spend most of our lives insulated from the dangers of nature. The war machine we discussed before? It's predicated on harnessing nature to beat the enemy. Even our more placid sanctuaries of civilization, our libraries and art collections, face an ongoing struggle to protect their contents from the ravages of nature. From huts to houses, we build walls and roofs to keep nature out.

This is supposedly a problem for true environmental enlightenment. Stalwart environmentalists sometimes complain that we are too dependent on technology, that our attitudes toward nature—and in many cases we really are just talking about an attitude adjustment—would be much better if we could just get out and experience nature for what it is, as it is, without all these conditioning technologies. The Discovery Channel's absurdist reality television show *Naked and Afraid* takes this position to an extreme. Every episode plops two enthusiastic survivalists in a wild, harsh environment without a stitch of clothing. Almost always, the naked survivalists shift rapidly from enthusiastic joy to tearful misery and exhaustion as they face the rigors of heat, cold, rain, wind, insects, darkness, and hunger. We also hear a related call from the environmental education community. We're too removed from nature, these people say. We need to bring the kids out from the cities and expose them to the wilderness. Or, alternatively, we also sometimes hear it from the prepper or survivalist community: We need to prepare ourselves to live on limited resources, without all these gadgets! We would never know how to handle ourselves in nature should our civilization collapse.

Other environmentalists bemoan the wide or collective impact that these air-conditioner technologies have on the environment. On a grand scale, they point out, deepwater drilling opens the door to massive ecosystem-destroying oil spills; feller bunchers—a contemporary incarnation of the Onceler's Super Axe Hacker—mow down a forest 20 times faster than a troupe of loggers with chain saws; massive earth movers make it possible for us to sheer off the top of mountains in search of their buried minerals. On a much smaller scale, in our own lives we routinely kill by a thousand cuts, participating in the so-called tragedy of the commons over and over again. While our conditioning technologies may improve each of our lives individually, they also create collective impacts we can only faintly experience. When we spray herbicides to control our weeds, when we flush our medications down the drain, when we discard our batteries in landfills, we don't act alone. All our neighbors do roughly the same, and all that waste collects in our rivers and streams.

Still other environmentalists charge that our air-conditioned lives deeply frame the ways in which we see nature. Our lives are ineluctably mediated by our technologies, so much so that we can almost not even fathom what it would be like to encounter nature on its own terms. We visit our national parks while sitting in the comfort of our cars, and the world outside takes on the feel of a movie or a visit to the zoo. We discover that there are gas and oil reserves locked in the shale beneath our homes, and the ground on which we live becomes less a place of tranquility and more a fossil fuel investment to be tapped when our kids go off to college.

More than this, these technologies shape our views of nature. The more we view the world through the lens of technology, the more the technology shapes our attitudes about nature. As we need oil and wood and fish and fowl to make our cities and our towns function, we lose sight of nature as replete with life, instead divvying it up into caches of standing reserves.[24] Our discussions of environmental matters gradually, quietly morph into a discussion of natural resource matters.

Unlike many of my compatriots in the environmental community, I don't share the same skepticism about science and technology. I'm not as concerned about our dependence on technology, the collective effects of living in a technological civilization, or even the way that technologies frame our experiences of the world—at least, not so much that I want to turn back the clock on technology and "simplify" our lives. (I am definitely concerned about these things. I'm just not ready to abandon most of our modern technologies in order to address them.) Many very advanced technologies have certainly improved and enriched our lives. Although we still have a depressing record on global poverty and illness, it is not clear that the world isn't better because of our technologies. At the same time, there is an important grain of truth in each of these critical stances.

Thing is, technologies are basically dumbwaiters. They function in our service, sometimes helping everyone live more productive and fulfilling lives, but sometimes also debasing humanity by generating a good bit of grief and devastation. In this respect the goodness or badness of any given technology is perhaps better conceived as a kind of social elevator. Technologies can take us where we want to go, elevating us to our potential, but they can also transport us into civilization's basement.

Where scientific discovery has removed the mystery of nature, and technology has empowered us to harness nature to serve our ends, we are only able to put these technologies to use if we already know what we want to achieve. A tarp is good for keeping our tents dry. A sleeping bag is good for keeping us warm. Bug spray is decent, though not great, at keeping the bugs away. All of this requires knowing what we're aiming at, what we hope to accomplish.

Science certainly helps demystify nature, and technology certainly helps control nature, but again, we're at a loss if we think that identifying objects in nature, or the functions of technology, can guide us. We cannot learn what we ought to do simply by improving our science and engineering. We have, for centuries, more or less thought otherwise. We've been relying on our bare intuitions to guide us. Varmints were widely viewed as a "problem," like weeds, so we deployed our scientists and technologists to eradicate them. But now that the retaining walls of civilization are bursting at capacity, we can't get away with their blanket annihilation, and we can't simply fall back on our intuitions about right and wrong.

And that raises one final point. Though there is a sense in which our technologies play an ineluctable mediating function in our relationship with nature, and there is a further sense in which we cannot escape their influence even when we shut them out entirely, there is a deep dark sense in which, despite our valiant attempts to keep it at bay, nature is not the "out there" that we often caricature it to be. Just as our technologies surround us and condition us, nature is unavoidably always at our feet, right where we are.

THE CRUELEST MISTRESS

An unnamed, unidentified, could-be-any family of three plays in the wooded park near its home in Anywhere, USA, near some anonymous big city, with dog and toddler in tow. It is a fine day for playing. The sun shines, a breeze caresses the hills, and the smell of autumn wafts peacefully through the branches. A small yellowish orb pokes its head out of the leaves, beckoning to the boy, "Come hither." Attracted to its contrast and shape, intrigued by its silky cap and thin stalk, the child wanders toward it. Upon reaching his goal, he squats, brushes away the leaves, and plucks a frail mushroom from the ground. The boy innocently examines his find, squeezing it between his diminutive fingers, looking closely at the small ring around the stalk, at the cup cradling the long stem, at the perfect gills on its underside. He crushes it between his index and forefinger, watching as bits of dust gather under his nails. Interest piqued, he places it in his mouth, rolling it around on his tongue. It tastes pleasant, like the salad mushrooms he has eaten at home. His parents talk among themselves, unaware of the horror that will soon befall their son.

The *Amanita phalloides*, better known to most as the death cap, will shortly wreak havoc on the boy's internal organs. Within a day he will exhibit symptoms not uncommon to most toddlers: a bit of colic, some diarrhea, and possibly even vomiting. His parents might not think much of his initial discomfort—accustomed, as most parents are, to this sort of thing. In two or three days his symptoms will disappear entirely. Life will return to normal. He will play with his toys, chase after balls, and poke around the house for the sundry artifacts of adulthood. For a day.

And then—then he may become slightly jaundiced, followed by delirium and seizures. His liver and kidneys will begin to fail as the mushroom toxin eats through both. The pressure in his head will grow intense. He will scream and cry as his brain tissue hemorrhages, as his pancreas ceases to function, as his heart twitches toward failure. If he is lucky, he will fall into a coma and die after only six days. If he is unlucky, his painful departure will stretch across two or more weeks.

Curiosity is a trait mostly reserved for cats, but in this case, small children can exhibit the same trait with far more devastating effects. Out of range of the watchful eye of their guardians, a single curious misstep can result in a gruesome end. For our phantom toddler, it was nearly too late the moment he put even a small bit of the mushroom in his mouth. Not so very long ago, before the days of liver transplantation, this child would have suffered certain demise. Good research and medicine has made it possible nowadays to save the child of this sad fate, although mortality rates for consumption of the deadly *Amanita* hover at around 30 percent, even in industrialized nations with a ready supply of livers for transplant. Mortality statistics are worse for small children, vulnerable as they are to both external threats and the pull of curiosity.

Thankfully, the story told here is only imaginary, though it is certainly not farfetched. In North America alone there are 10 to 20 death cap poisonings per year, many of which are attributable to toddler grazing. These low numbers, like the statistics for bear attacks, are as much an artifact of our good sense to stay the hell away from nature as they are a result of healthy eating. In Eastern Europe and Russia, where mushroom hunting is a national pastime as well as a route to sustenance, the number of deaths from poisoning is considerably higher. With the recent surge in transcontinental immigration, many Asian families in particular, accustomed to picking the delectable paddy straw mushroom, an unfortunate look-alike of the death cap, have learned the hard way that they are not to be eaten.

We talk, very often, as if one can enter into and exit out of nature, as if one wanders into the woods, like the Ives boy or me on my camping trips. We talk as if the hazards lurk only ever "out there," "in the wild." But we never really escape nature. Even in our womblike cities and our sprawling suburbs, nature is the medium in which we function. There really is no clear demarcation between the wild and the tame, between wilderness and city. Just as our conditioning technologies permeate all aspects of nature, so too does nature permeate all aspects of our lives. It's everywhere: in our yards, our homes, our bathrooms, our guts, our pantries. It's inescapable.

Which raises one of the most curious but prevalent arguments in the green arsenal—against the plasticity of Twinkies and Ho-Hos, against the fluorescent halo reflecting on the shine of prepackaged and processed foods, against the genetic modification of our vegetables—regarding the "naturalness" or organicity of certain foodstuffs. Natural and organic foodstuffs are, allegedly, produced "out there," in "nature," away from the corruption of chemicals and artifice. The idea, of course, is that natural products are good and healthy, whereas artificial products are not so. Invariably, as each semester rolls along, I am stuck with several students who push this line. They are insistent that nature will provide for us much more efficiently and healthily that we can possibly provide for ourselves. Many of them are local food activists, persuaded that locally grown, indigenous varieties of plants and animals are healthier, tastier, and better. The same argument, much to my amusement, also often appears in discussions about the legalization of marijuana. "It's natural," the proponents of legalization protest, implying that whatever is natural could not possibly do any harm. Many of us, myself sometimes included, are inclined to believe that because something is natural, it is good.

Michael Pollan and the gurus of the new green are as much to blame for this view that "natural equals better" as anybody else—though to be fair, Pollan, advocate of healthy and sustainable eating, author of the widely read *In Defense of Food* and *The Omnivore's Dilemma*, is reasonably tempered in his view. Again, I don't object to eating natural foods—I agree that we all should eat food, mostly plants, and Pollan's work is an encouraging step in the right direction—but it is important to note the tendency to emphasize the good in the natural while downplaying the bad.

I went on my first wild mushroom foray while I was living and doing research in St. Petersburg, Russia. The activity is a national pastime, and almost all Russians have reasonably strong knowledge of local mushroom varieties. In my case, I had to trust in my friends to guide me toward the choice edibles and steer me away from the toxic varieties. Pollan details a similar interest in mushroom hunting in his writing on the dilemmas facing omnivores. But he neglects at least one terrifying aspect of mushroom hunting: if you pick the wrong one, you're in big trouble.

As should be plain to all, the fact that something is natural is not at all an indication that it will bring about desirable or even good outcomes. So is the case with our death cap, as well as with other amanitas and toxic mushrooms that go by equally menacing names: the destroying angel, the devil's finger, the jack o'lantern.

Naturalness is not synonymous with healthiness. Jasmine berries are natural, yet they will cause such intense bloating that you'll have sympathetic daydreams of Violet Beauregard, the bubblegum fanatic who inflated to the size of an enormous blueberry in Willy Wonka's chocolate factory. Hyacinth and daffodils are beautiful ornamentals, but if you mistake their bulbs for onions, you may be pushing up more than daisies. Don't sate your desires by eating those seductive red mistletoe berries, either; they're the kiss of death. Two castor beans will knock an adult to his knees with a fatal dose of ricin. Yew will kill you. While you're at it, avoid azaleas, rhododendrons, and laurels. All deadly.

It is even true that the allegedly safest and most natural foods carry hidden dangers. In early 2009, the Ohio-based King Nut Company was singled out as the distributor of the organic peanut butter that was responsible for Salmonella poisoning in at least 410 people, several of whom died. The peanut butter incident was one of a string of regularly occurring outbreaks in which packaged food products, sometimes trumpeted as organic and healthy, were found to be contaminated with bacterial pathogens. Even with our controls and interventions and our attempts to lock out the bad, there ain't no such thing as a hazard-free lunch.

As I've said, it's not the risk or potential danger of harming ourselves that is my concern here. I'm exploring the downsides of nature only to make a point, which is that whether nature is good or bad is an extremely complicated claim, made all the more difficult by the fact that nature is everywhere, and it is, at the same time, ineluctably conditioned by all the human things we do. Some of nature is good, yes; but some of it is very bad. The goodness or badness of nature can't be hung on some feature intrinsic to nature, or some relationship that we have to it, nor can it be attributed to some clear set of benefits that nature allegedly gives to us. There is no magic dividing line between the natural and the unnatural. The natural is everywhere. The artificial is everywhere. We can't escape this.

One might rightly ask: Now what? Where do we look for guidance? What are we supposed to do to be green?

If we accept the view that environmentalism is about loving nature or finding value in nature, there is no apparent answer to these questions. Some environmentalists say one thing while others say another. We are perched on a knife's edge between doing harm to ourselves and doing harm to nature. We could go on for quite a long time hashing out the merits of either the intrinsic or the extrinsic value approach. We might even just throw up our hands and call it a day. The standard view that environmentalism requires identifying nature's value would appear, at least upon examination, to be untenable and unacceptable, leaving us with no clear direction. Seems to me, though, that rather than abandoning the environmentalist project altogether, we should call into question the standard view.

MOMMY DEAREST

Nature brings us both the good and the bad. It can be good for us. It can be bad for us. It may be good in itself. It may be good *for* itself. Given our technological entanglement, it's hard these days even to say with certainty what *nature* is. We do not know how nature lives or where it works. We do not have access to its machinery or magic. For the Ives family, the loss of their son was no doubt an unparalleled catastrophe—even though, scientifically speaking, the danger of bear attacks is well-known. For soldiers in Vietnam, nature was a catastrophe amplifier—even though, morally speaking, the ravages of war are also known. For parents of our imaginary toddler and for victims of the King Nut salmonella outbreak, nature was an unexpected catastrophe. Natural catastrophes come in all shapes and sizes. They don't discriminate among populations. They don't discriminate according to wealth or manner of dress. They are avoidable only in circumstances where we wrest control of nature and take steps to ensure that they are avoided.

After all, nature doesn't take the time to care whether it's *your* son or daughter between its jaws. Nature just acts according to law, to instinct, to need. If a carnivore needs food, it will seek out delicious morsels. If an outcropping must fall, it will slide to the bottom. If a volcano must erupt, it'll do so with magnificent force. The earth and its creatures function according to laws entirely indifferent to the concerns of humans. When we let our guard down with nature, we die. This is a problem for all of us, and it has been a core problem since time immemorial.

Which all raises this peculiar conundrum: Why are so many environmentalists intent on emphasizing the good side of nature? Nature can be pretty awful. Wouldn't it be best to just admit as much?

I certainly think so. Would doing so leave us in a lurch? I think not.

Look more closely at the various tales I've recounted here. Stories such as these remind us of our own involvement in the outcome, of our potential to do otherwise—to prevent, thwart, or interfere with an outcome. When we rub up against nature, we're not just passive recipients of the world around us. We are participants in nature. The Ives family sought more information and wondered what they might have done differently if they had had that information at the time. The US military in Vietnam sought to use information and technology to alter nature to strategic advantage. At any moment we can choose to accept the curveballs that nature throws our way or choose to take a different path. We are unique in that we have the power of choice. Sometimes, of course, we simply don't have a choice, as in the Ives case or the case of the toddler, since nature unravels too quickly for us to react. But many times we have information about where nature is going, about what we can do to divert its trajectory, about what we might do to yield a different outcome.

What we have seen in this chapter is an overarching negative side of nature. Nature sends her handmaidens into our campgrounds in the middle of the night to suck the lifeblood out of our children and our pets; she shields the enemy during wartime, providing solace and nourishment to those who intend us harm while also amplifying the effects of our weapons to successfully achieve our ends; she entices us with the prospect of her temperament, remaining still and harmless until we blindly taste what she has to offer; and she is relentless even in the face of our protective technologies, returning to rear her head no matter the extent to which we suppress her nefarious ways. No wonder, then, that we should characterize our relationship to nature as a colossal struggle.

Our age is essentially a tragic age, but we have for years refused to take it tragically. We have learned that catastrophes happen, that they may happen again. Robust scientific studies warn of impending ruin, yet we continue building up new little habitats, finding new little hopes, oblivious to our falling skies. It is rather hard work, this determining what to do. How are we to know what's right, where to go? There is no smooth road into the future. Are we to divide our world into two columns, the good and the bad, to sort it out according to what suits us? Are we to choose the lesser of two evils—man or nature—as is so often assumed?

It would be easy to fall victim to the argument that evidently persuaded the flower children, that we should make love, not war, which perhaps inspired legions of later environmentalists to believe that the core theoretical challenge was to get everyone else to appreciate nature. It would be easy to think that the clearest route to environmental peace and tranquility is the one that identifies the good in nature, that extols the wonderful and the magnificent. It would be easy, that is, if one could turn one's back on all this darkness, if one could ignore the incredible bad that accompanies nature's magnificent good.

As it happens, figuring out the right thing to do isn't so easy.

4

The Wild and the Wicked

In which the author (1) introduces two comparison cases and sets up the central moral distinction of the book, (2) characterizes outcomes as end states of the world, (3) models decisions about what to do in terms of decision trees, and (4) proposes a distinction between actions and events.

CRASH BOOM

The day after Christmas is for many Westerners a day of serenity and regrouping: relief from the flurry that has characterized the weeks leading up to the arrival of Santa Claus. Children bounce happily from room to room, leaving a trail of toys and empty boxes in their wake. Parents lounge in their new pajamas and robes, sinking sip-wise into steaming cups of hot chocolate or coffee. Grandparents discuss whether to make a gargantuan breakfast or to glide on the high from the still-digesting dinner of the day before. The chaos of the previous few weeks finally hits a lull, and all can enjoy, for perhaps the first time, one another.

The Boxing Day Tsunami of 2004 therefore came as an unwelcome surprise to millions of locals and vacationers on the otherwise halcyon dawn that broke peacefully in Southeast Asia. An early morning undersea earthquake forced an upheaval of the ocean floor, displacing millions of gallons of water and causing an oceanic shock wave that, from space, must have looked like a rippling belch in an earth-shaped pond. Torrents of water gushed furiously onto the shores of Thailand, India, Indonesia, Sri Lanka, and several other spots around the Indian Ocean. Over a span of a mere 15 minutes, hundreds of thousands of people were slammed against buildings, into trees, through glass, under cars, and face first into the sandy loam of the shoreline. Lovers walking on the beach, children playing in the sand, parents basking in the sun, elderly couples enjoying their families—nobody was immune. Virtually all who stood haplessly in the path of the mammoth wave were hurtled about, the crashing force of nature tossing them like rag dolls. In the end, approximately 230,000 people were drowned, dragged out to sea, or crushed to death by debris.

Pictures and video captured during the event only faintly convey the horror that the victims endured during those fifteen minutes. Even the most graphic video shows the wave as a relatively gentle 5- to 10-foot surge of water pushing over walls and lounge chairs, spilling into resort and hotel lobbies, moving cars and boats around like toys in a bathtub. To get a true sense of the magnitude of the calamity, one would have to interview each and every person affected, relive the event through the eyes of every victim, have the experience of being slammed into docks over and over again, bones cracking, skin ripping, lungs filling, organs collapsing—the heaviness and horror of death catalogued in an audio encyclopedia of more than a quarter million entries.

Though estimates vary, the approximate energy of the undersea earthquake that spawned the tsunami was equivalent to between 250 and 800 megatons of TNT. For reference, this means that the earthquake was 20,000 to 60,000 times more powerful than the atomic bomb that the United States dropped on Hiroshima near the end of World War II. These numbers are unfathomable to the average person, but if you can imagine as many Hiroshimas detonating underneath the ocean, lifting millions of acre feet of water in an instant, and dumping them like the Teton Dam on the banks of your favorite sunning spot, you may get some idea of the magnitude of forces at play. If you're still having trouble imaging those figures, try dropping a gallon of milk on your foot from waist-high and multiplying by 1 billion. That's about 8.8 kilotons.

The bombing of Hiroshima is generally regarded as one of the most devastating man-made events in human history. It ranks right up there with the top 10 natural disasters, of which the Boxing Day Tsunami is only the 9th most deadly. (The 1931 Yellow River flood in China is estimated to have killed between 1 and 2 million people, but it played out over days, not hours.) When the innocent-sounding “Little Boy” atomic bomb unleashed 13 kilotons of energy over the center of Hiroshima on August 6, 1945, more than 70,000 people were incinerated instantly.

As days stretched into months and months stretched into years, hair fell out, thyroids bloated, and fatalities climbed. Within three months of the bombing, roughly 70,000 more people died of illness caused by burns, radiation poisoning, or scarce medical resources. By 1950, the death toll rose to 200,000 from this single event. All told, it is reasonable to believe that still many others died from causes only indirectly related to the bombing—lives lost and missed by the rigid criteria of finicky actuaries. More important, these numbers do not include the tens of thousands of survivors who were injured and deformed in the blast, the numbers of birth defects, the cancers, the animal deaths, the costs of reconstruction, and the interrupted lives. Some bomb-affected persons (*hibakusha*) live on today, carrying a radioactive legacy bequeathed to them by an orchestra of scientists, politicians, and soldiers.

Both the Boxing Day Tsunami and the bombing of Hiroshima caused unfathomable heartache. Both events caused costly devastation. Both events were the result of forces the magnitude of which cannot adequately be captured in a single paragraph, not to mention a thousand granite memorials. Only one event, however, caused regret, guilt, anger, and condemnation, as well as approval, whistles, cheers, and relief. There’s a reason for this, captured nicely by the distinction that many philosophers make between the *good* and the *right*.

Let’s put it this way.

The tsunami was an act of nature, the result of an enormous undersea earthquake that originated near Sumatra. It ranks in the world’s top 10 natural disasters, distinguished, as such, by their unpreventability. Physical forces, outside the control of humankind, met one another head-on that day, and almost no amount of planning could have prevented the ensuing chaos. True, a warning system may have saved lives. True, breakwaters farther offshore may have prevented damage to the resorts. True, a more coordinated and organized response effort may have spared people their children or their limbs. But everything else being equal, the tsunami was an unpreventable nightmare. Given that it was an act of nature, it can only really be described on a spectrum of good or bad, fantastic or terrible. It would be odd to say that it was somehow *wrong*.

The bombing of Hiroshima, in sharp contrast, was not unpreventable at all. Hiroshima was an act of human beings, the result of countless hours of planning, coordination, secrecy, not to mention, millions of dollars. It ranks in the world's top 10 most deadly man-made events. After considerable deliberation and discussion, many influential and not so influential human beings, working in coordination with one another, ripped open the back door of physical law and disintegrated 64 kilograms of uranium in a millisecond. Mother Nature does not take lightly to the instantaneous disintegration of her heavy metals, and thus unleashed furious energy in response, destroying a city and taking with her all life. True, this heated response was made possible by the unforgiving grace of Mother Nature, as no coordinated effort of human energy could have wreaked such hell, but in the end, the bombing of Hiroshima was an act of humans.

Given that it was an act of humans, it can be understood not only as simply good or bad but also as right or wrong. When humans take action, there's something far more complex going on.

FREEZE-FRAME: STATES OF THE WORLD

What is interesting about the Boxing Day Tsunami and the bombing of Hiroshima is that they both eventuated in effectively the same outcome: tremendous numbers of people were killed, and many more people were made miserable by the event. It is a fair bet that if a misery meter had been keyed into the emotions of all human beings on the earth, the needle would have gone off the charts in both cases. To read just a single story of a survivor of either calamity makes one want to plug one's ears and slam one's eyes shut, to flit one's tongue over one's lips with clattery "la-la-la" sounds, as if noise could possibly squelch out the pain that accompanies such horrific experiences.

If I've done my job here, you'll agree with me that Hiroshima and the tsunami were both very bad events from the standpoint of misery. When I say this, I'm not suggesting that they were equivalently morally bad. I'm just talking about *experientially* bad, particularly for those millions who were either directly or indirectly affected by the events. Having one's skin melt is presumably quite a gruesome way to go, as is being slammed up against cars and ocean floors and drowned to death. It is also true that both of the two events were the result of incredible forces, the magnitudes of which take significant mental gymnastics to understand. They caused billions of dollars of damage, leveled cities and homes, and restructured societies. Speaking strictly in terms of subjective misery and objective harm, the events were pretty awful.

Given that both of these calamities rank equally high on the physical and emotional devastation scale, we tend to want to make mental comparisons. I think the natural tendency in these cases is to compare each by appeal to what was more or less horrific. So, for instance, we might say that exploding instantaneously in a ball of fire would be far better than drowning; or we might say that the long-term effects of radiation outweigh the relatively minor costs of water-borne disease resulting from the tsunami's contamination of the water supply. We can also compare aspects internal to the events themselves. We might imagine that those in the epicenter of Hiroshima may have had it better than those on the periphery, who died a slow and agonizing death. We might also say that it would have been better to be knocked unconscious first and then dragged out to sea, rather than being dragged out to sea and kicking endlessly in a futile attempt to return to shore.

• • •

When I was younger, my friends and I used to engage in comparative discussions such as these all the time, usually in reference to the impending nuclear war that we all anticipated: "I'd prefer to die instantly," one of us would say, "I don't want to feel anything." Another of us might disagree, claiming that it would be nice to be able to say good-bye, to know what was happening. Most would certainly argue that less pain is better, but there was always one masochistic kid who would argue that it would be better to die in a painful way, to be able to experience your own death. Usually my friends and I would carry on like this at our bus stop, waiting for the cheese wagon to pick us up and ferry us off to school. It was an endless game, requiring imagination, contemplation, argumentation, and qualification.

What we were doing was making value claims, assigning the value *good* or *bad*, or *better* or *worse*, to some aspect of whatever event intrigued us. Such assignments of value were relatively straightforward for measurable quantities—like megatons of TNT, dollars of damage, lives lost, children orphaned—but a little less straightforward for the subjective experiences of humans and animals. We couldn't possibly have ready access to the thoughts and experiences of all involved—gone, as they were, to a watery or fiery grave. But because we did have experiences with related events—grabbing a scalding cast-iron pan, struggling for oxygen while holding our breath at the bottom of a pool—we assumed we could at least approximate the horror of the incident.

Approximating horrible subjective experiences is no easy undertaking. Yet almost all of us do it on a regular basis: children do it, adults do it, and political pundits do it. It's one way we make sense of the world. It's so commonplace that we may not even be aware that we're doing so until it backfires on us. Sometimes, for instance, these attempts at approximating horrific or joyous experiences interfere with the most personal moments of our lives. We often talk this way to console those who have experienced a crisis, saying that we "can only imagine" how awful it must have been. When we do so, our responses may have the undesired affect of upsetting rather than consoling those who have actually been injured in or witnessed related catastrophes. "No, you cannot imagine," they may respond sullenly, "it was a horror so horrible that even the most productive imagination could not possibly imagine."

When we make such estimations of the badness of events, we're doing something else as well. What we're doing is taking a step back from the event itself and freeze-framing it. Putting it in a freeze-frame helps us get a handle on what we're valuing. Like film critics or armchair quarterbacks sitting in front of the tube, we pause the action and offer a value assessment about what has just happened, what is happening, or what will happen. "Wouldn't it be awesome ['good'] if Butch and Sundance could make it out of Bolivia?"

So even though we are sometimes talking about relatively private and personal experiences—like what it must be like to have a child's fingers slip away as he is swept out to sea—we treat these values as though they are assignable to *states of the world*: either facts about the actual world or, and this is the tricky part, facts about some possible world that could have been but is not. We say, "It could have been a whole lot worse," for instance, by which we mean that the Germans could have gotten the bomb before we did or that the wave could have struck around noon, when thousands more were on the beaches. In those cases, we're imagining a state of the world that does not exist, that never existed, but well might have. We imagine some alternative universe, some alternative possibility, and assign a value to *that* state of the universe.

But perhaps I should pause here. Maybe your first thought is that I'm wrong about the misery caused by Hiroshima, that the atom bomb put an end to a bloody world war and surely, in the grand scheme of things, this *prevented* misery. For all the misery that it caused, perhaps this was offset by all the good that was done. It is said that the world cheered when the bombs were dropped on Hiroshima and Nagasaki. Who can forget the famous images of ticker-tape parades in Times Square? Allied supporters were surely overjoyed to have their boys coming home. Survivors of Pearl Harbor were probably stunned into silence, but felt that some justice had been done. Even some Japanese citizens and soldiers, no doubt, were relieved to finally see the end of the war. The *bomb* did that, despite the grief that it rained on Hiroshima.[1]

I think, as I've said, that this kind of comparison is bus-stop guessing game. You and I can pass time until the bus comes arguing about what makes one event better or worse. I could offer a similar counterargument about the tsunami: that from the standpoint of the world, the Boxing Day Tsunami stabilized the tectonic sheets for the near future, perhaps avoiding a further, worse calamity, and it also functioned as a preventive call to action, inspiring governments around the world to institute a warning system that could prevent further loss of life in an even greater disaster.

It is important to see not that one state of the world is better or worse, but that debate about better and worse states of the world is commonplace in our everyday discussions. If your objection to what I've said about Hiroshima is that the war was ended, then you are doing exactly this right now. You're claiming that there is some alternative state of the world that could have been but wasn't—the state of the world in which the war would have continued until it ended in another, potentially more devastating, way. I'm not seeking to offer a condemnation or approbation of the bombing of Hiroshima, only to point out that we regularly and often make objective value claims about events, states of affairs—states of the world. This is a natural way of thinking for us.

We make these kinds of claims about states of the world and assign values—good or bad, better or worse—to them all the time. Any self-respecting businessperson will admit that state-of-the-world talk is how companies and small businesses make decisions: by weighing the outcome of investments. The same goes for many other areas of life as well. We do it in sports, as when a "bad" call moves the runner to the next base. ("If only that umpire had been paying attention!") We do it with weather events, as some years have comparably worse snowfall than others. ("If it snows as little this year as it did last year, the ski resorts will have to cut the season short. That would stink.") We do it with traffic jams, hospital stays, vacations, concerts, automobile accidents, and almost all other possible events. The knee-jerk response that some tragedy "could have been worse" is used the world over to comfort those who have had a bad experience. Virtually anybody who has ever suffered an accidental loss, who has ever gotten into trouble or made a stupid decision, knows what it is to say that a given state of the world is worse than it should be or could have been.

This value talk—this talk of the good (good and bad, better and worse) is virtually inextricable from state-of-the-world talk: how many dollars were lost, how much pain there was, how much happiness there will be, how many people have been affected, what certain stakeholders might want, how many are dead, how many are injured, how many square miles were affected, and so on, causing vertigo in even the most enthusiastic actuary. When we freeze-frame the world, we are inevitably drawn to describe it not only as it is, as a state that can be understood according to some description, but also as a state that has some comparative value as well: good or bad, better or worse. As we've seen, such discussions about states of the world are not limited to the external world. They also apply to the mental and experiential world, to those who have been made miserable or happy.

It is probably fair to say, for instance, that the world would be a *better* world had the tsunami not happened. We can imagine this, and often do. A child who was left stranded on the beach as her mother was pulled out to sea must believe that the world is now suddenly much worse than it was when she woke up that morning. She may ponder for years how the world would have been different had she and her mother been just 100 yards up the shore, or had she waited just a few more hours to urge her mother to go out to play. So here's where this gets interesting.

THE LITTLE ENGINE THAT SHOULD: DECISION JUNCTURES

States of the world branch out before us as so many hopeful possibilities, and they fall behind us in the vast and fractured wreckage of what was and what could have been. When we make a choice, when we choose to act, we simultaneously open up a plethora of new possible worlds and leave a trail of unrealized worlds in our wake.

Philosophers sometimes talk about these as *decision junctures*, the moments at which we must move from deliberation to action. Normal people use a wide range of less technical expressions to describe these junctures: the moment at which we "pull the trigger," "take the first step," "switch tracks," "step up to the plate," or "make the plunge." Our language to describe this moment is richly metaphorical. It conveys the idea that, like a train conductor, over the course of one's day, over the course of one's life, one faces a series of junctions, at which time one could either pull the lever left and steer the train in one direction or pull the lever right and steer the train in the other direction. We can even draw little pictures of train tracks (sometimes called *decision trees*) to help us understand our decision better.

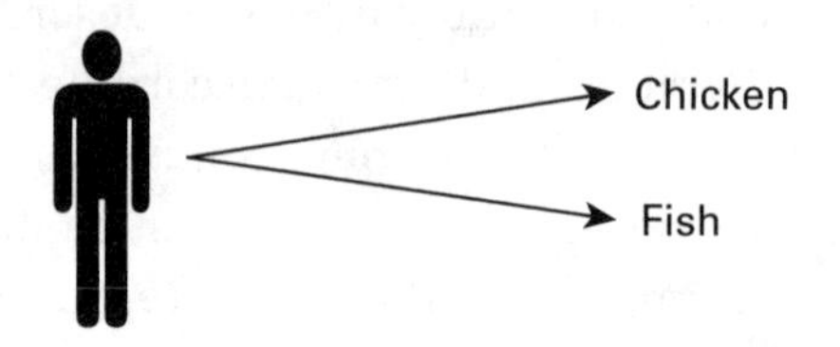

The idea is reasonably intuitive, but for clarity's sake, imagine a trip to the grocery store. You have $10 and must pick an entrée for your evening meal. And so you must decide: What should you buy? Such decisions about what to buy are generally not fraught with much moral weight. It's just dinner, after all. But there are important dimensions to such decisions that may otherwise go unnoticed. One of those dimensions is that your decision forecloses on further possibilities. If you buy chicken, you cannot buy fish. If you buy fish, you cannot buy chicken. With only $10, you must choose one. Not only this, but if you opt for fish you will not be able to serve your magnificent chicken Kiev, which your guests loved the last time. If you opt for chicken, your guests will not benefit from the curative powers of foods rich in omega-3 fatty acids.

But then, the other dimension is that, even though you are foreclosing on one set of possibilities, your decision opens up a brand new landscape of possibilities. If you buy fish, you will be able to make ceviche or sushi or poach it in wine. If you buy chicken, you will be able to barbecue it or pan fry it or serve it Kiev-style. As one set of possibilities is foreclosed upon, a new set of possibilities blooms.

This is a decision juncture. Every decision juncture collapses a set of possible states of the world and opens up a range of possible new states of the world. These decision junctures specify the moments at which deliberation ceases, at which your body—arms, legs, hands, feet—becomes responsive to your mind and shapes the real material world into a form slightly more congruent with what you intend (or, in some cases, do not intend). It's a magical moment, in a way, because your mind otherwise has little access to the world of your body; and yet, you somehow transform your thoughts, deliberations, and choices—your reasons—into action. Your mind and your faculties of perception function as a filter through which states of the world pass, through which you, using your body, can shape the world.

Just as anyone who has ever experienced a bad state of the world knows what it is to wonder what "could have been," any of us can look back on actions we have taken and evaluate whether we made the right choices. It's curious, but this appears to be an integral feature of how we understand ourselves. We are pleased with our decisions if they turn out well or better than we expected. Maybe our decision for chicken was great, and our guests cannot stop raving about the dinner. And we regret decisions we've made, or disapprove of decisions that others have made, if those decisions turn out poorly. We may regret having chosen fish, since we should have remembered that Sally doesn't like fish. The state of the world that we created—a world in which Sally had nothing to eat—was uncomfortable, not what we'd hoped for, bad.

Sometimes we regret our decisions, curiously, even when they don't clearly relate to what went wrong. If we are in a bus accident or a plane crash, we may blame ourselves for having gotten on the bus or having booked that flight. Many wives and husbands of those killed in the terrorist attacks of September 11, 2001, have described the remorse they felt for letting their spouses go to work. Their choices plainly had nothing to do with the horrible events of that day, yet most of us can understand their reactions. Many people who were directly affected by the attacks have grappled not only with the world as it could have been but also with things that they could have done differently—their own personal choices.

The point here is that like state-of-the-world talk, decision-juncture talk is distinctive of our human experience. We freeze-frame and evaluate states of the world partly because we recognize alternative possible states of the world, but also because we recognize the possibility of humans—either ourselves or others—doing otherwise and creating a different world. Even in cases where decision junctures do not *causally* lead to a given outcome, we are given to analyses that understand states of the world *in terms of* the decisions that we *can* make. What if we hadn't vacationed in

Phuket? What if General Tibbets had refused to fly the Enola Gay? What if, like the girl on the beach, we had been 100 yards farther up shore?

Coupled with states of the world, decision junctures make it possible for us to talk about what we *ought* to do—what we *should* have done or *could* have done differently—not just what *is* the case. They acknowledge freedom. If we didn't face decision junctures, we would be little more than helpless billiard balls crashing up against one another, shuttled around by the laws of physics. We could only ever describe our behavior in the third person or in the passive voice. "It happened today that John was at the store." "Tina was nourished by chicken Kiev." "Joe was made angry by the absence of fish." Groceries, in short, would just "end up" in our carts. There would be no quandary about what to do.

Decision junctures, characterized both by your *decision* to pull that lever, and the actual movement of your limbs in the pulling of that lever, eventuate in states of the world. You will be having fish for dinner. We generally call this whole process of making a decision and altering the state of the world an *action* or an *act*. Freely chosen actions, in turn, make it conceivable to talk about morality. You can choose fish or chicken, just as much as you can choose whether to harm or help another. Some actions are just actions, but others have a moral dimension.

Of course, you can't choose everything. You often don't get to choose your options, for instance; the world gives most of them to you. If the store has no chicken, you have no choice for chicken, unless you seek out another store. If there is no grocery store, your only choice is to find another possible source for dinner. Decision junctures drawn as decision trees can mask these external considerations. They are descriptive only of our personal options, and if we're not careful we may forget to include in our description of our actions all of the limiting factors that go into making our decision.

What is deceptive about decision junctures, then, is that it can be easy to think that one is only ever making a choice between end states of the world: between good or bad, better or worse, chicken or fish. It is tempting to think that all one need do is overlay one's values on the resulting states of the world and then choose which is better. "Like fish better than chicken? Turn left." But many philosophers think that there's even more to it than that. They think that the complexity of human actions isn't adequately—and can never be completely—captured by the characterization of decisions as simple choices between states of the world. Instead, decision junctures must also include some consideration of the principles and constraints that guide a person to choose those particular possibilities.

Don't get it? Return to our shopping trip. When you enter a grocery store, you bring with you not only a set of values and desires but also a set of principles and constraints that define the rails on which your train can run. To see this, just change the scenario a bit.

Suppose you are in a foreign bazaar with a much broader range of choices. Suppose that a particular store offers not just fish and chicken but also more exotic treats like snails, caviar, rattlesnake, gopher, nutria, and, as a special, the meat of orphaned human children. Jonathan Swift parodied this last menu item in *A Modest Proposal*, in which he suggested it as a solution to all the vagabond and poor children in Ireland. He attested then, and we can only assume that the same still holds, that "a young healthy child well nursed is at a year old a most delicious, nourishing, and wholesome food, whether stewed, roasted, baked, or boiled." So our fancy foreign bazaar will take Swift up on his proposal and also sell the meat of these children. Here's another decision tree perhaps a bit more descriptive of this scenario:

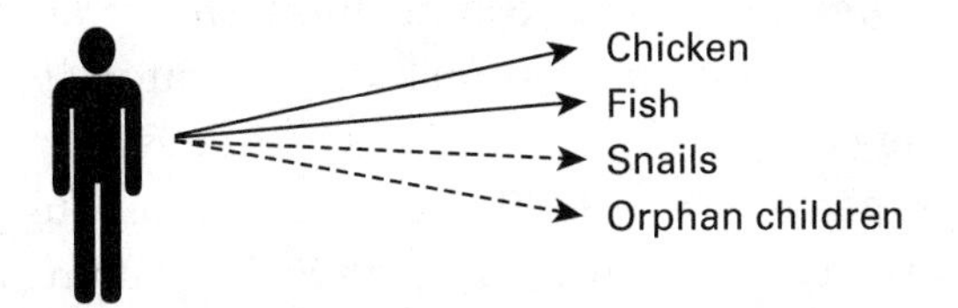

For most of us, orphan children are not a viable option for dinner. For the sake of this argument I'll suppose that you are of such upstanding moral fiber that you would not, on moral grounds, eat an orphan child. Bully for you. But suppose also that even though you personally would love a treat of escargot, you're not sure if your guests will be so enthusiastic. We might rightly then leave both options off the list. We might say that in this case you have only two real choices: chicken or fish. We might even draw a diagram like the one we saw earlier, perhaps this time with dotted lines to indicate that snails and orphan children are an option, but not a *live option*. The tricky part here is that your reasons for keeping snails off the list differ dramatically in nature from your reasons for keeping orphan children off. Your reasons for avoiding snails are based on what your guests may dislike. Your reasons for avoiding orphan children are based on moral restrictions.

It gets even trickier if you consider that there are ways of including these options in your decision tree that seem almost, though not completely, plausible. If you had more information about your guests' preferences, you might well include snails on your list of culinary possibilities. Perhaps they have had snails, but they have not had *your butter and garlic* snails, which you think will turn them into snail aficionados. What I hope you sense, however, is that under *no circumstances* would the flesh of orphan children present itself as a viable dinner option. No matter, even if it were free or prepared by the finest chefs in the world to be tastier and more succulent than the purest and highest-grade veal. Jonathan Swift was hoping that you'd see it this way. So there is a different kind of restriction going on with orphan meat than with snail meat. The difference in the nature of these reasons is central.

If you approach your decisions strictly by looking at states of the world, you will undoubtedly miss the wide range of reasons (principles, preferences, and constraints) that might otherwise guide your actions—reasons that include your duties to others, your hidden preferences, and your own or your guests' restrictions on treating others as mere objects. If you do include among your choices options that are not really options, what you'll be doing is collapsing out reasons.

The reasons that you crowd out are the internalized parameters of your decisions. They guide your decision options and characterize your choices, ultimately shaping your actions. It may be that you refuse to purchase human meat, animal meat, or anything cooked blackened and with Cajun spices. It may be that you have responsibilities to do something but can't bring yourself to do it. The internalized parameters contain a much broader spectrum of reasons for your actions, some of which guide you in your valuations, some of which constrain you in your choices, and some of which urge you to do something that you might not otherwise want to do. Sometimes your reasons are attached to things you like and don't like. Your distaste for snails may be of this variety. At other times your reasons are attached to restrictions that you take to be fundamental constraints on human behavior: legal prohibitions or the requirements of morality. Your refusal to eat orphan children may be of this variety.

RUBBER, MEET ROAD: ACTIONS AND EVENTS

On July 16, 1945, in a remote location near a small town called Alamogordo, the glassy white sands of New Mexico's desert bore witness to an explosion never before felt on earth. This was Trinity, the proving ground for physicists and mathematicians from Los Alamos National Labs and the detonation destination of the world's first plutonium bomb. J. Robert Oppenheimer, primary developer of this weapon and director of the Manhattan Project, watched with a mixture of relief, wonder, and concern as the plutonium bomb ushered in the atomic age. He would later famously say of his thoughts that day:

> We knew the world would not be the same. A few people laughed, a few people cried, most people were silent. I remembered the line from the Hindu scripture, the Bhagavad-Gita. Vishnu is trying to persuade the Prince that he should do his duty and to impress him takes on his multi-armed form and says, "Now, I am become Death, the destroyer of worlds." I suppose we all thought that, one way or another.[2]

It's a familiar thought to anyone who has ever accepted responsibility for an action. Andrei Sakharov, Soviet physicist and developer of the Teller-Ulam hydrogen bomb—a bomb hundreds of times more powerful than Little Boy—had many of the same feelings as

Oppenheimer, but he was motivated to do even more. Shortly after the first H-bomb test, Sakharov renounced his work in physics to become one of the world's great pacifists. He worked tirelessly to prevent the proliferation of nuclear weapons. Up until his death in 1989, he was famous for trying to undo the state of the world that he had helped bring about. Some chalked this up to guilt, and that may have been part of it, but it was also an explicit recognition of his own power in transforming the world.

By contrast, the Boxing Day Tsunami did not result in widespread disapproval, but in an outpouring of compassion for the victims. Within hours much of the world began organizing. Scores of nongovernmental organizations flew to affected areas and immediately began providing food, water, and medical support. In just a few weeks relief organizations gathered a staggering $10 billion in aid, much of which, amazingly, came not from governments but from private donors. Most of the rest of the world was treated to a cavalcade of media—first about the causes of the tsunami; then, when they ran out of stories about that event, about other extreme acts of nature. This raises the following point:

It is a curious fact about the English vernacular that we refer to events like the Boxing Day Tsunami as *acts of nature*. This is fantastically misleading. Tsunamis are no more *acts* than the whirring of your computer's hard drive is an act. Tsunamis are extremely complicated physical and geological events—billiard balls crashing up against one another. They are the result of forces. As we have seen above, actions are generally characterized by an *actor*, a decision maker, the human force behind the event. Events are just, well, states of the world, characterizable primarily by a shift from one state of the world to a new state of the world.

Perhaps our use of this vernacular is a hangover from more religious times. We sometimes refer to weather events as *acts of God*, by which we mean that God has dunked his hand into fishbowl earth and swirled it around, wreaking havoc on the humans below. Indeed, even today airlines use the phrase *acts of God* to describe weather events that interrupt plane schedules. Following the devastation of Hurricane Katrina, many people were quoted in news reports as saying that their lives had been turned upside down by

God's will. Almost all natural disasters can be spoken of in this way and often still are. As the Enlightenment demystified religious language, however, the term *nature* stepped in as pinch-hitter.

But there's a deep problem with this idea. When we talk about acts of nature in this secular era, there's a real tendency to see natural events in the same light as human actions. Yet acts of nature are clearly not the same sorts of acts as the acts of humans. It's unusually deceptive to talk about *acts* in this way, much as it is for me to refer to nature as the gendered and anthropomorphic Mother Nature. I sometimes use this terminology for a reason, of course, and I do it somewhat tongue-in-cheek, but many people continue to call certain events acts of nature, as though nature were a deliberate force. Most often when we talk about acts, we speak in terms of deliberate actions, in terms of decision junctures, as we have seen above: I acted to open the door; the rapist acted upon his victim with aggression; the company acted to stave off losses. If we confuse events in nature with actions, then it would appear that we can attribute responsibility to nature, just as we attribute it to humans. But clearly that's preposterous. Nature doesn't have reasons for her events. Humans do.

Here's why this matters. Some people believe that the way to make ethical decisions is by appealing to what will be better or worse once an action has been taken. To do this, they propose that you should look at your suite of action options and evaluate the goodness or badness that those actions might bring about. You've heard this kind of appeal before. Take Hiroshima, for instance. It is often said that the bombing of Hiroshima ended the War. That seems like a pretty good outcome; seems like somebody made the right decision. To a lesser extent, you too may think along these lines when you're making decisions about, say, what to buy for dinner. If you think like this, you may be subscribing to the view that the right action is the action that brings about the most good. Fine for decisions about dinner, but maybe not the whole story for ethical decisions. These are the people I will later be calling *do-gooders*.

In contrast, some people believe that the way to make ethical decisions is by appealing to what is right. They propose that you should look not at the events caused by your actions—since human actions, as well as natural events, change states of the world—but instead at the actions themselves. Those who think along these lines take into account myriad considerations about intention, motive, principle, constraint—the full gamut of *reasons*. So, for instance, they are likely to fault a person for intending to harm even if no harm is done. They might argue that having intent to kill is wrong even if one does not in fact kill. More important, however, they may have deep-running concerns about what counts as an action and what does not count as an action. This is because they think that one can be accused of wrongdoing based on the *reasons that guide one's actions* regardless of the outcome and, in some cases, regardless even of one's intent. You would be wrong to purchase orphan meat for your guests even if orphans are as delicious as Jonathan Swift says they are. That's why the distinction between the tsunami and Hiroshima is so important. The tsunami was an event, where the bombing of Hiroshima was a coordinated human action, replete with *reasons* that we can evaluate. For the sake of this distinction, let's call people who think this way the *right-reasons* crowd.

It may seem that one need not draw this division so starkly, that there is no clear need to distinguish between the do-gooders and the right-reasons group. We should, obviously, bring about a better state of the world, and we should, presumably, also do so for the right reasons. But the distinction is actually pretty important. Depending on which view you assume, you may be driven to radically different conclusions about the right course of action. One view may instruct you to allow for the killing of innocents in order to bring about an end to a war, whereas the other may suggest that some techniques for bringing about peace are strictly off the table. The same view may instruct you that it is sometimes permissible to lie (such as to avoid hurting someone's feelings), where the other may place stringent restrictions on lying (because you are obligated in principle to tell the truth).

Though both views are often in agreement about the majority of cases, the rationale that underwrites these decisions can generate quite a bit of disagreement on more marginal cases. The do-gooder, of course, relies on a fairly compelling intuition: if anything would make one act better than another, it must be that it creates or produces a better state of affairs. What could be wrong with this, with making people happier or increasing net benefits in the world? There are many strong arguments in favor of the do-gooder approach, and much of the time they can be quite persuasive within moral theory. It makes a lot of intuitive sense to say that when it comes to acting ethically, what matters the most is whether the world is made better. Critically, however, the do-gooders are saying that as long as you bring about this better state of the world, it doesn't matter what your reasons are.

The right-reasons crowd will disagree. They will suggest that your reasons matter as well, that we have to consider not only outcomes but also your reasons. They'll get beaten up for this view, because it looks to a lot of critics that this will result in some counterintuitive positions—for instance, that in order to do right, you may have to do harm. Ethical actions that actually make the world worse? How could that be?

It is vitally important to see that when we're choosing courses of action we absolutely must determine what states of the world will result from those actions. We simply can't make decisions without looking at outcomes, because actions, if they are anything, are also always events. So this isn't really in question. Attention to outcomes holds for both the do-gooder position and the right-reasons position. What *is* in question, however, is whether having the right reasons matters at all. I argue that it does, that those actions that are right are also reasonable and can be *justified*—although I have a very specific notion of justification in mind. I'll return to that in the latter half of this book.

My guiding thought here is that we in the environmental community face a somewhat difficult challenge that is complicated by the way in which appeals to a better world have been implemented in policy making. In justifying their positions, many environmental policy makers make appeals to a confusing suite of ends: to promote nature or to advance human welfare or to grow the economy or to conserve ecosystem benefits or to fulfill any of the other values that we touched on in chapter 2. These appeals, compelling as they may be for some, are not always faithful to the more careful thoughts of the founding do-gooders: Jeremy Bentham, John Stuart Mill, Henry Sidgwick, and more contemporary figures like Peter Singer and R. M. Hare. We environmentalists are left with basically two options: either to firm up the do-gooder position, as it is represented in the environmental community, so that it better accords with its theoretical roots; or to look to other branches of moral theory for guidance. For most of the rest of this book I will spill a lot of ink taking this latter position and leave the former defense to my colleagues in other branches of philosophy.

The way to get a grip on this issue is to embrace the cleavage between actions and events, because it's here that we find the root of the matter: reasons. For most people, the distinction between human actions and natural events is intuitive. It is as straightforward as the distinction between killing someone and allowing them to die from natural causes. It guides many of our public policies, but particularly our legal code. One can be found guilty of murder or manslaughter and be held liable for premeditated crimes or accidental harm, but typically not found guilty for accidents involving falling rocks, wild animal attacks, or the existence of cancer. Circumstances that may change the liability of an individual—say, cases of negligence—are those in which it can be demonstrated that the individual was somehow complicit in some way, perhaps through negligence, or perhaps by not taking proper precautions when one had the jurisdiction and the obligation to do so.

But some people want to deny that the distinction between actions and events is morally relevant. They think that it doesn't matter if you refrain from eating orphan meat because you think it will be disgusting. All that matters, in their view, is that you refrain from eating it. "No harm, no foul," some say. But as we can see, this can result in some odd conclusions. Think of the case of the toddler discussed in the last chapter.

Suppose that that we are back in the park with our family from chapter 3. A toddler is on a hike with his parents, just as before, but instead of accidentally stumbling upon the *Amanita phalloides* (death cap) mushroom, a stranger intentionally hands him the mushroom, knowing full well that the child will die an agonizing death if he ingests it. Here we have a scenario not much different in outcome from the first scenario—the child dies—but very different in terms of human involvement. Let's call this the Case of the Poisoning Stranger.

I feel comfortable saying that this case is unequivocally bad, and in many ways much worse than the case in which the child eats the mushroom by accident. It is tempting to nod one's head here in agreement. To say, "Yes, the original case was bad, but the case with the stranger is *extra bad*." I'm sometimes inclined to that kind of thinking, but I'm not sure it's right to be. So let's examine this thought.

What, then, is *extra bad* about this event? It would appear that there is an extra consideration in the Case of the Poisoning Stranger: a very bad person acted deliberately to feed an innocent boy a lethal mushroom. So there's the "extra bad" feature that does the work: deliberate poisoning. It is also probably fair to say, however, that the act was bad primarily because the child died and not because the stranger acted deliberately. The stranger's act was made considerably worse by his *success* at killing the child. We can even make this into a kind of equation.

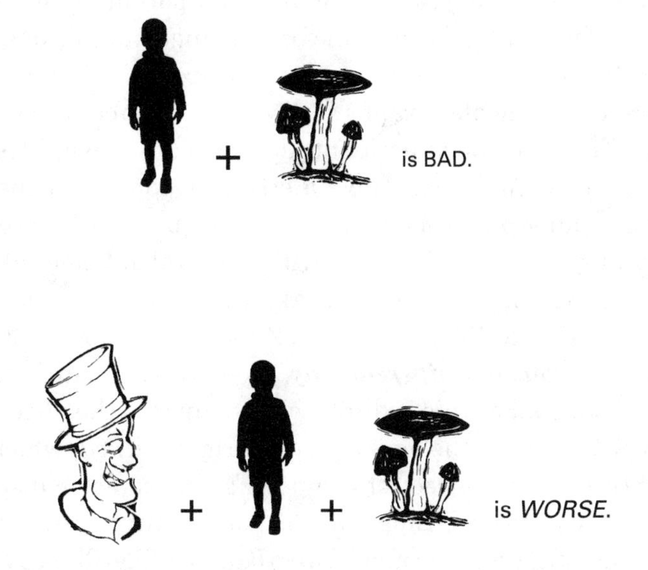

If you agree that scenario B is worse than scenario A, then surely you will also agree that such considerations about the intent of the stranger are not equivalent in badness to the true outcome of the event, which is the death of a young boy. And if *that's* true, then you will surely also agree that if *two* little boys had been wandering innocently through the park, and both had accidentally eaten poisonous mushrooms resulting in their deaths, then this must be worse than the case in which the poisoning stranger kills just one boy with the mushroom. In other words, you would agree that the death of two little boys by accidental mushroom poisoning is worse than the death of one little boy by intentional mushroom poisoning.

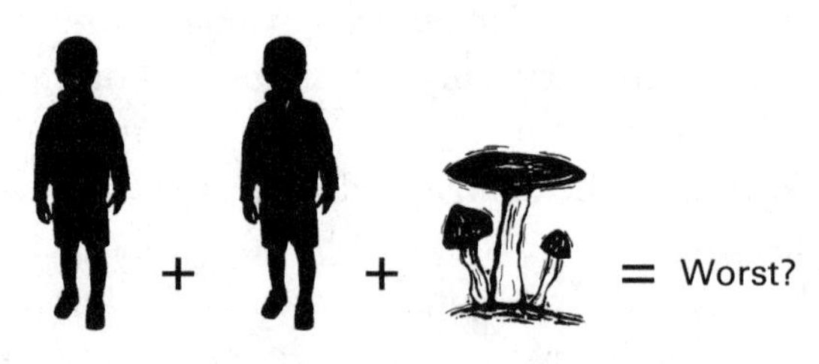

I don't know about you, but my intuition is that this conclusion is wrong. My intuition is that although it would be tragic to have two toddlers die from accidental mushroom poisoning, it would be *bad in a different way* for a malicious stranger to kill one boy by deliberately feeding him a poisonous mushroom. When I consider the two cases, the noble seraphs in my mind trumpet the following objection: "Yes, two boys accidentally poisoning themselves with mushrooms is awful, but to have a stranger pop out of the weeds and give a toddler a poisonous mushroom is just *evil*." Since I don't think morality has anything to do with evil, which I generally associate with some kind of supernatural possession—"I'm speaking to the person inside of Regan now"—we need another way of understanding the "*bad in a different way*" element associated with this state of the world. For this, I think, we can dip into the virtue literature and characterize the action as downright *wicked*, which to my mind captures some aspect of the actor's character or disposition rather than his essence. Elphaba was wicked but not evil. Freddie Kruger? The other way around. Über-baddies like Ramsay Bolton challenge this distinction a bit, but hey, work with me.

The bigger point here may be a bit obscure, but it should become evident if you consider your own intuitions. If your intuition is like mine (that scenario C isn't clearly worse than scenario B but is bad in a different way), then I suspect you'll agree that we should really find a way to describe this difference. If, however, you think that scenario C clearly *is* worse than scenario B, then you must at least agree that in order to hold this view you must attribute some quantity of additional badness to the actions of the poisoning stranger. And if you characterize the actions of the poisoning stranger as additionally bad, then you also need to offer an account of how, and to what extent, human intentionality factors into your moral calculus. In my view, wickedness does not just add an extra touch of badness to any given event. Rather, wickedness, and any moral evaluation of action, involves an assessment that is orthogonal to a mere evaluation of badness.

Again, I'm not really trying to get you to agree with me that one scenario is worse than the other. I'm primarily trying to get you to see that when we talk like this, we seek to compare one state of the world with another state of the world. In this case, we inevitably factor out many of the central considerations that are distinctive of our moral landscape. Just as with our decision junctures—our constraints against eating orphan children—we are doing more than comparing state of the world B, in which one stranger lethally poisons one boy, with state of the world C, in which two young boys accidentally and naively poison themselves by picking and eating poisonous mushrooms.

You may feel the same discomfort with characterizing the poisoning stranger's intent as one bad heaped on top of another set of similar bads. If so, you're rubbing up against the friction that I think should guide you. What we're doing in this case is evaluating states of the world, and that characterization seems ... well ... deficient. We flatten out the texture of right and wrong by transforming it into a claim about states of the world. If you can sense this deficiency, then you're well on your way to understanding the controversy over the good and the right.

Would your feelings about the guilt of the stranger change if you learned that he didn't know that the mushroom was toxic but instead was aiming to show the child something beautiful or curious in nature? What if you came to find out that the stranger was crazy? Or what if the stranger had been sent on a mission by God? Suppose the stranger's family was being held captive by a mobster who told the stranger that the only way to rescue his family would be to feed the toxic mushroom to the toddler. Would you still think of the situation in the same way?

My suspicion is that your feelings about the guilt of the stranger *would* change—you would want to know more, you would want to know *why*, *why* that stranger did what he did. You would want to know motive, intent, aim, extenuating circumstances—*reasons*. You would want to *understand* what happened, and it would be very different from the way that you would want to understand why a toddler accidentally ate a poisonous mushroom. You wouldn't have the same question *why* in the accidental cases. Sure, you might question the competence of the parents, such as their decision to let their son wander, but you would not question the reason for the event. There is no reason. No justification can be offered. These things just "happen."

In short, merely assessing states of the world does not open the door to reasons, and it seems to me that it's these reasons that we're after when we talk about morality and ethics.

It is my view that to uncover the right reasons we have to examine full-blown actions and not simply states of the world. We make a grave error if we dive into deep ethical waters without acknowledging the reasoning process and the wide range of reasons that make up our decision junctures. If we think that we can pass judgment on an action simply by appealing to the end state of the world (or even the expected end state of the world), we become little more than moral mathematicians, assigning values of good and bad, better and worse, to possible outcomes—like kids waiting for the bus. The human project of ethics is reduced, then, to a simple actuarial project of crunching the numbers. If that's the case, the mathematics of morality must be a funny math indeed.

"We're all faced throughout our lives with agonizing decisions, moral choices. Some are on a grand scale, most of these choices are on lesser points. But we define ourselves by the choices we have made. We are, in fact, the sum total of our choices. Events unfold so unpredictably, so unfairly, human happiness does not seem to be included in the design of creation. It is only we, with our capacity to love, who give meaning to the indifferent universe. And yet most human beings seem to have the ability to keep trying to find joy from simple things, like their family, their work, and the hope that future generations might understand more."

—Professor Levy, *Crimes and Misdemeanors*

5

Control Freak

In which the author turns to an examination of intentional action, (1) exploring first the reasons that we make choices, (2) examining two different ways of understanding actions, (3) distinguishing between willing and wanting, (4) dissecting the various reasons, and (5) characterizing the will in terms of principles and habits.

BOUND AND DETERMINED

Once upon a time near a moiling stream there hiked two hikers, a man and a woman—Susan and Robert, let's call them—who had left their sleepy town in Maine to raise funds for the mentally ill by hiking the Appalachian Trail. Educators and do-gooders both, the pair had arrived at the Wapiti Shelter, a small wooden lean-to, to eat and gab and rest their weary heels amid the dark hollows before their long journey would begin all over again the next day.[1] The stream by which they pitched their camp went by the ominous name of Dismal Creek; and a dismally good name it was, for less than 24 hours after they dropped their packs by that shelter, they would lay dead in their sleeping bags, buried under loose dirt and piles of leaves, Robert with three bullets in his skull and Susan pierced and punctured through the back with a long nail and a knife.[2]

Were this only a fairy tale, it would surely have ended on a happier note. Susan and Robert would have woken the following morn, cooked breakfast on their stoves, and set along their merry way, without a care but for the mosquitoes, the poison ivy, and the occasional twisted ankle. As it happens, however, what our fairy tale lacks in "fairy," it makes up for in "'cautionary." The time upon which this tale occurred was not a fictitious but a specific date in the summer of 1981, and our fellow travelers were figments of neither the Brothers Grimm nor Mother Goose, but ill-fated real-worlders whose lives were tragically cut short by the workings of one Randall Lee Smith, a local madman who had grown up in the nearby town of Pearisburg, Virginia. The pair was discovered ten days after an extensive manhunt led right back to the bloody floorboards of the Wapiti Shelter—or perhaps it was the Wapitu Shelter, nobody really knows, for the "u" on the sign had grown so mossy and illegible that it made sense only to guess at the final vowel.

By all accounts this was a terrible occurrence. It was awful in many ways, but certainly all the more awful because it appears to have been entirely unnecessary. Like our Case of the Poisoning Stranger, Smith popped out of the bushes and wrought his rage on Susan Ramsay and Robert Mountford Jr. in a way that virtually belies explanation. Smith could have chosen a different course of action. He might simply have stayed at home to watch late-night television or hung back eating popcorn and catfish. Had he done so, the two hikers would likely still be alive today.

For a time, the Appalachian Trail Murders, as they later came to be called, generated concern among the hiking community. The safety and serenity of the woods was cast under a shroud of fear, and those who love the trail and the sanctuary that trees provide knew not whether civilization could be counted on to stay in its proper place. To stanch this concern, officials moved into action, just as they did after the Ives bear attack. "Violent crime is rare," said Brian B. King, spokesman for the Appalachian Trail Conservancy. "You have more of a chance getting hurt driving to the trail in your car than you do on the trail."[3] He is correct, of course. The list of murderous attacks on hikers and campers is shorter than the list of car accidents. Or bear attacks, for that matter. But the consolations of bureaucrats are cold comfort for the skittish.

Statistics alone cannot mollify the jitters of aspiring outdoor recreationalists. The public also needs clarification—a reassuring explanation to help offset one random murder from other everyday occurrences. What did Smith want? Why did he want it? What were his intentions? How did this happen? Like the dinosaur hunters who centuries ago explained their buried findings in terms of a catastrophic global deluge instigated by a malevolent god, investigators began the arduous task of explaining the sordid state of affairs and, more directly, Smith's sordid state of mind.

As with many explanatory undertakings, the first place to look was the perpetrator's festering cauldron of desires. What were his motives? Perhaps Smith was angry, or he desired sex, or he wanted money. It would be helpful if it could at least be said that he intended to kill, but even this much is difficult. Lacking a mind-reading device, investigators had little clue what Smith intended. Instead, they tried to work backward from what they knew, piecing the puzzle together and extrapolating further facts. Where there is motive, there is possible intent.

Smith struggled with Susan but shot Robert in cold blood. It would appear that he needed to get Robert out of the way so that he could rape Susan. He desired sexual gratification. Sex is a pretty powerful motivator, so most can understand this. Tom Lawson, the investigating officer, suspected exactly this: that Smith's motive in killing Susan was primarily sexual, whereas his murder of Robert was motivated by "self-preservation"—presumably to allow him to get away with his crime.

The identification of a motive helps us understand whether Smith's action was premeditated or impulsive and, ultimately, whether it was intentional or accidental. The state of mind of the criminal is also critical to our assignment of blame. It is so critical, in fact, that it is taken up in our legal code. The law recognizes a range of mitigating circumstances for any crime: insanity, postpartum depression, unintentional harm, diminished capacity, self-defense, the year-and-a-day rule, and so on. Most people find the logic reasonably compelling: morally wrong acts that are committed intentionally, with forethought or with malice, are quite a bit more reprehensible and problematic than those that are mere accidents.

Explanations and characterizations of a murderer's motives also help us keep the murderer at arm's length, positioning him as unlike us, but also making sense of whatever it was that drove him to act in his criminal way. Sometimes, but not always, explaining his behavior in terms of his motives even undercuts the framework by which we assign blame and guilt. Explanatory motivations only may or may not serve as an exculpating or mitigating reason. Their moral upshot is an open question. The challenge for investigators, armchair psychologists, and philosophers lies in understanding the difference between a motivation and a reason.

In the case of the Appalachian Trail murders, Smith was found guilty of second-degree murder and sentenced to 30 years in prison in a plea bargain, only half of which he would serve. Second-degree murder, if you're not aware, lies halfway between first and third degree murders. It designates an intentional act of murder that is not premeditated or that is committed as a crime of passion. Smith essentially traded 15 years of freedom for the murder of two people, all of which may or may not be enough to get your blood boiling, depending on how you feel about the facts of this case.

In the last chapter I discussed decisions, an essential component of choice. I proposed that our choices about what to do aren't adequately characterized by the narrow path-based approach of decision trees. Now I'd like to turn to another dimension of choosing, which I think should further call into question the effectiveness of thinking narrowly about choices. Namely, I'd like to dive deeper into actions, to get a clear sense of what we really should be assessing to identify what is right or wrong in any given action.

THE DRIVER AND THE DRIVEN

The characterization of human behavior in terms of the drives or motivations is, in fact, a fairly prevalent way of thinking about things. None other than the great Sigmund Freud generally claims the scepter as the psychologist most responsible for plumbing the depths of the drives, but I'll hardly do him the disservice of laying all the blame at his feet. Many other important thinkers have also characterized human reason as tied directly to the drives. Philosopher David Hume famously claimed reason as a slave of the passions. Jeremy Bentham begins his seminal work on utilitarianism, *The Principles of Morals and Legislation*, with one of my favorite openings in philosophy: "Nature has placed mankind under the governance of two sovereign masters, pain and pleasure. It is for them alone to point out what we ought to do as well as to determine what we shall do. On the one hand, the standard of right and wrong, on the other the chain of causes and effects, are fastened to their throne."[4] Friedrich Nietzsche understood morality and value as intertwined with the innermost fears and desires of individuals. Even Plato painted a picture of the individual soul as divided into three functional components that closely presage Freud's id, ego, and superego.

Nevertheless, Freud and his cadre of followers—basically, any and almost all who collectively make up the modern psychology establishment—have had a profound impact on our thinking about human action. The various schools of psychology—which are by no means limited to psychoanalysis but also include behaviorism, functionalism, gestalt theory, structuralism, and cognitivism—seek to understand human motivation the way a surgeon might understand the circulatory system or a physicist might understand the thermodynamic properties of trinitrotoluene. Each school of psychology treats human motivation like an attribute of the subject, to be dissected, examined, and put back together again. This conception of human behavior, in turn, influences our ideas about responsibility, obligation, guilt, and blame. In essence, it dramatically shapes our sense of what it is to make a choice.

Speaking very generally, the psychological answer to the question of what happened—whether in a murder case or regarding what one bought for dinner—unveils beliefs and desires as explanatory of a person's motives. These beliefs and desires are determined by backward-looking reconstructions of events: the boy ate the mushroom because he was hungry and believed the mushroom to be food, or the boy did so because he was curious and believed the mushroom to be safe. The answer credits the beliefs and desires with an explanation of the event and purports to offer insight into the behavior of human beings.

Psychology is great for this sort of thing. It helps detectives solve murders. It helps nondetectives make sense of madness. It helps us understand our children when they act out, our parents when they are lost and confused, and even ourselves when we do things for which we don't have clear reasons. Judging from the recent spate of books on this or that road to happiness, it can even offer guidance on the straightest path to the good life.

Psychologists have recently tried to claim dominion over ethics as well. Authors like Sam Harris (in his heavily criticized *The Moral Landscape*) have caught the attention of philosophers and psychologists alike. Harris proposes, enticingly, that morality can be approached productively through the standards of the empirical sciences. It's an appealing angle: science has given us clarity on innumerable other questions, so why not also morality? Harris seems to think that we can get a grip on morality simply by identifying the states of affairs that will get us to the point of greatest well-being. I strongly disagree with this view, as do many philosophers, but I can only touch on one criticism here.

For all of their promise, psychological explanations are really just a species of explanation, not unlike the accounts of the dinosaurs offered by Buckland and Mantell. As a philosopher, I find such attempts at explanation unnecessarily limiting. They not only account for a relatively one-dimensional way of thinking about a person's actions, but they also crowd out other explanations that otherwise give credibility to the reasoning powers of the person. In effect, they reduce the reasons that may have moved a person to act down to a collection of internalized drives.

For instance, one way of thinking about the Appalachian Trail killings is to suppose that something drove Smith to commit these acts. Maybe he harbored perverse desires—to rape or shed blood. Maybe he had false or crazy beliefs—that he was being persecuted by spiders, or that Robert was a demon. Who can say what the origin of these desires and beliefs was? Maybe it was his environment, maybe his genetic composition, or maybe something deep in his past. We don't know. Whatever the case, the idea that we might gain insight into his actions if we just plumb his psychological depths depends vitally on an attitude toward actions that treats them like events. Z happened because A, B, and C coalesced to cause Z to happen.

These motivational explanations can indeed go a long way in helping us understand the long chain of events. Notice how easy it is to slide between *cause* language and *motive* language. It makes sense. If we want to understand the reasons people act the way they do, we would do well to get inside their heads—to probe their psychology, their psychological states, their frames of mind, just as before we looked at states of the world. The problem is that if we answer the motive question in causal terms, we're halfway out of the moral ballpark. It's not long before these attempts at clarification turn to characterizations of the perpetrator as a brute, an animal, a monster—scratching, clawing, biting, grabbing his way out of his primal desires.

In the Appalachian Trail murders, for instance, legions of evaluators sought to explain Smith's behavior through a variety of familiar lenses, any one of which would serve perfectly well to explain his actions. He was a pervert. He was scared. He was a monster. There's a case to make for each of these positions. Good investigators will cover all angles, seeking the best explanation.

But most important, every one of these psychological explanations paints Smith as subject to some set of forces, internal or otherwise, that swirled and sloshed and mixed to squint his eye and move his arm to raise the knife or pull the trigger. All these explanations treat Smith as a brute, driven by some force outside his control. Mind you, I'm not saying he wasn't a brute; but he was a brute only metaphorically speaking. The problem is that the moment we invoke brute talk, morality talk goes flying out the window.

Consider this: A puppy will pull a pie off the table because it is driven by its desire to eat food. A bear will upend picnic baskets in search of tasty morsels. A horse will chase carrots and fear sticks. These animals do these things to satisfy their needs and desires. There is little self-direction apart from a general description of their behavior as oriented toward the simple satisfaction of desires. The objects of their desire are dangled before them by outside forces—they needn't be pie obviously; they could as easily be a cheese sandwich or a bag of cookies or a dirty sock—and the animals are just pushed along—driven, as it were, into the food option.

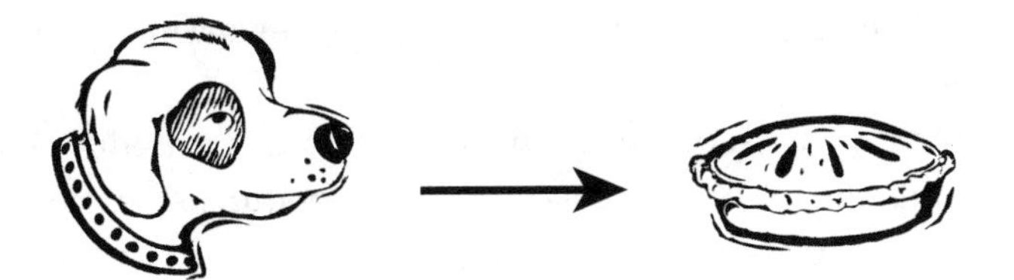

This all happens with varying degrees of intelligent involvement, of course, as many ethologists will attest. Ants and caterpillars are driven to acquire food differently than black bears and walruses, but the basic idea is the same. Critters seize opportunities to acquire what they want and need. They are driven by their impulses, inclinations, and desires.

But this, it seems to me, isn't entirely what people are up to when they make decisions about what to do. They don't simply see something that will satisfy their desires and take it like a pie from the table. When they do, we hold them accountable for having done wrong, particularly if there is a reason not to treat the objects of their desires like pieces of pie.

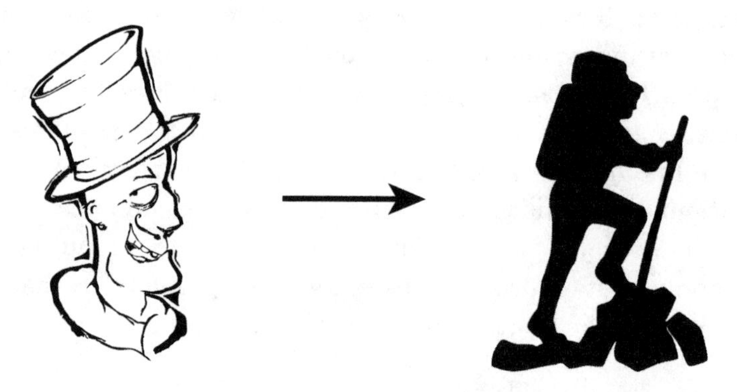

It is true, of course, that people aren't objects to be used and disposed of. Nor are orphan children objects to be killed, poisoned, or eaten. It is also vital that as moral agents—people with *agency*, the capacity to steer our behavior according to the principles of right and wrong—we understand the difference between women and objects or children and food. But when talking about wrongdoing, we should not let loose our grip on the person taking the action. This idea of agency is important to morality. We do not hold bears morally responsible for rooting around our campsites at night. We do, however, hold people like Smith morally responsible. We do this because people are morally blameworthy whereas animals are not.

So far as I can see, assigning blame to a wrongdoer—or conversely, doing the right thing as a person—requires acknowledging that in some circumstances people are not driven by these external motivating factors—food, sex, survival, lollipops—but are instead sitting in the driver's seat and can, if they see reason to, step on the brake. They can be motivated by some other source of reasons. Without some acknowledgment of this, it is very hard to blame wrongdoers. "He's just acting like the pig that he is. He's hardwired that way." It's hard to blame a guy if he's little more than a pig.

What lies at the base of this notion of moral blameworthiness, then, isn't strictly speaking the objects of our desires, nor whether and how objects have been treated, but the extent to which individuals are self-directed enough to act otherwise, to act according to principles that respect the variety of reasons for not simply taking things when we want them. Historically, philosophers have proposed that sometimes our desires for things are wired directly into our impulses, as when we are feeling particularly animalistic, but that sometimes we take action for other, bigger reasons.

What Smith exhibited, more than simply a capacity to do otherwise, and what makes this crime so heinous and wicked, was a capacity to reason, however inadequate it may have been. So what if he was drawn to Susan? So what if he was an angry, malice-filled nut? As a human being, it is a fairly safe bet that he maintained at least some capacity, however minimal, to understand murder as wrong. He could have taken steps to prevent himself from doing what he did—lashing himself to a mast, perhaps, like Odysseus. But he didn't. He failed to. He failed at being human. And it is for this that we fault him.

The lens of objectivity is thus deceptive when it comes to matters of ethics. Since we cannot get at blame by way of these objective explanations, even if we glare down from a divine perch like the Reverend William Buckland DD FRS, or if we peel back layers of skin and peek inside someone's skull like ole Professor Freud, we must look elsewhere for some sense of how to assign blame. The philosophical approach is to don the glasses of reason, to try to make sense of the reasons that can be said to reside in a self-directed individual. After all, it is the self-directed individual—the individual with the steering capacity to direct his or her life in one way as opposed to another—not the driven individual—the individual who is battered to and fro by the forces of nature, by the caprice of his or her desires—at whose feet we lay the blame.

The question that we should therefore be seeking an answer to is not necessarily what drove Smith to act in this way, but what makes him and his actions so terrifyingly wicked? Psychologists have several possible replies. Some may say that it was in his nature. Some may say that it was nurture, and the whole charade can go on for a very long time. But do these sorts of answers actually ever get to the true nature of his wickedness? I think they don't, and that is because we're really asking a very different question. We're not asking what forces have made Smith wicked. We're asking under what conditions we should characterize his actions as wicked.

Actions are distinguished from events primarily insofar as actions are taken by agents, people who act for *normative*, as well as brute, reasons. When we learn of some terrible event, or sometimes even when we learn of a wonderful event, our natural response is to characterize the event as either good or bad. When we learn of some action, in contrast, our tendency is to subject it to a different sort of evaluation: that of moral praise and blame, approbation and disapprobation. In the Appalachian Trail murders we see this sort of praise and blame directed not at Smith's animalistic wants and desires—in the eventlike features of the action—but in the inability of Smith to keep these animalistic wants and desires under control.

OUTSIDE FORCES

The gruesome coda to the Appalachian Trail episode takes us right back to where it all began, at the Wapiti Shelter, 28 years after Randall Lee Smith plea-bargained his way into a reduced sentence of second-degree murder, 12 years after he was released on parole, and 4 days before he would die in prison from injuries sustained in a car crash.

For those 12 free years, Smith had lived a relatively uneventful life, under the constant and watchful eye of his parole officer. He took menial jobs. He stayed close to home. And then one day in 2008 he disappeared, violating his parole. Unbeknownst to his handlers, Smith wandered back onto the Appalachian Trail, not to be seen for days. Missing-person flyers went up. People began to worry. The police initiated a search for him.

That same weekend Scott Johnston and Sean Farmer had gone on a fishing trip, as they had many times in the past, oblivious to the macabre history of the location that they would innocently pick for their camp. In the middle of the afternoon, not long after catching a few fish for dinner, Scott and Sean were approached by a gaunt stranger who introduced himself as Ricky Williams, something eminently human beaconing from his eyes. They invited him to dine with them. Hungrily, he obliged.

They cooked. They chatted. Williams told tall tales of working with NASA, of his education at Virginia Tech, and when all was done, the moon was high, and the crickets were chirping their fine staccato songs, Randall Smith, not Ricky Williams, pulled out a 0.22-caliber pistol, pointed it at Sean's head, and shot him in the temple. He then coldly turned to Scott and pulled the trigger a second time, hitting Scott in the neck. Continuing, he again turned the gun on Sean, this time striking him in the chest. And then once again he turned back to Scott, piercing his neck a second time. Back and forth, four times, as an appetizer, a prelude, to what would follow. He was so close to his victims that they had powder burns around their gunshot wounds. One bleeding from the head, the other from the neck, the two friends scrambled frantically into the shadowy woods to escape, leaning on each other for support.

Amazingly, Sean Farmer managed to stumble to his truck and climb inside, on-again-off-again fountains of blood pulsing from the finger-size holes in his head. He had no clue whether Scott was still alive, but the momentary silence permitted him an instant to collect his scattered thoughts. Either luck or stupidity had left Smith with an empty gun, which was barely enough time for Sean to shout to Scott to hop in the truck, start the engine, and begin the arduous task of driving down the darkened mountain, sticky and glistening in the moonlight. Smith stood alone, back at the site where it all began, unhappily holding his warm gun.

So much for our earlier speculations about what might have driven Smith to kill.

The rest of the story is an unbelievable intermingling of neighborly goodness and technological intervention. There's a farmhouse, a scared resident, bleeding on the porch, a medevac helicopter, a police chase, and an eventual car crash that would give Smith the fatal wound that would end his life in prison. Astonishingly, both Sean Farmer and Scott Johnston managed to survive this brutal ordeal, albeit barely.

Washington Post journalist Wil Haygood would later describe Smith as "a killer seemingly without motive. A man who wouldn't explain. A man who emerged from a life of misery to suddenly strike back at the light around him."[5] Haygood rightly notes that we fear people who kill for no apparent reason much more than we fear mere killers.

Lacking a clear motive, your natural inclination may be to turn to environmental factors that influenced Smith's personality. There's actually quite a bit in his past to explain his behavior. Randall Lee Smith was a habitual liar. He went by the nickname "Lyin' Randall." Says Sherman Smith (no relation) "The tales he'd tell, and—they were just kinda outrageous, and—and I—I think he—he expect people to believe him, but—I don't think anybody believed any of it."[6]

Randall Lee Smith was an only child raised by a single mother who barely held down a job in the laundry room at Giles Memorial Hospital in Virginia. He was described variously by those who knew him as a loner. As a child his mother dressed him in girls' clothing, seemingly for no clear reason and without explanation.[7] He would randomly walk the Appalachian Trail on his own, sometimes disappearing for days. Any of these explanations is a perfectly plausible clue to the constellation of forces that drove Smith to commit these acts. Perhaps his environment was such that it created the monster that he came to be. Perhaps he was the product of a long line of odd birds. Perhaps he had a chemical imbalance.

Yet the important observation here is not what led Smith to kill. He was a killer, whether plain and simple or complicated and obtuse. The important observation is *what we're doing when we ask these questions*. And what we often do is confuse questions about his actions—the acts of killing—with questions about the events themselves —like a tsunami or a rock falling on your head. Where on one hand we evaluate a perpetrator as morally blameworthy or guilty, on the other hand we explain his actions much as we might explain the Ives attack. We're searching for an explanation of the causal chain of events when what we should really be doing is asking whether the reasons that motivated him (whether they be the brutish reasons of his animal nature or the more complicated normative reasons of his human nature) are enough to justify or excuse his behavior. So what we want to know is whether, and to what extent, the reasons that guided him are justifiable. That is how we arrive at an assessment of the wickedness of his actions.

Instead we frequently attend to questions about actions as though they are slightly more complicated events. We treat Smith like an object, subject to external forces not of his own origin: biochemical imbalances, environmental pressures, early childhood programming, springs and levers in his head, and so on. Pairing this with good or bad states of affairs gives us a default response to the justifiability (or unjustifiability) of his actions. But it also makes it extraordinarily confusing and hard to understand what is going on when we make these important observations about motives.

Notice how readily these motivational explanations bleed over into the apparent excusal of an act. Many will adamantly deny causal psychological explanations out of concern that such explanations somehow exculpate the perpetrators of the wicked action. "How can you say he was a product of his environment?" they may protest. "He knew exactly what he was doing!" It is my view that these sorts of objections are classic instances of missing the point; we can sensibly talk about someone's motives or drives without introducing any moral evaluation or excusal of the person's actions. To do so, however, we have to be considerably more cautious.

THE MIDNIGHT CUPCAKE THIEF: WILLING AND WANTING

Long ago and for many years philosophers talked about something they called the *Will*, by which they essentially isolated the rational capacity of a person. It's not really commonplace anymore to speak of this Will, since it evokes images of a soul, what's meant by it is somewhat vague, and—to speakers of English, at least—it seems strange to capitalize a noun in the middle of a sentence. Nevertheless, I find the idea helpful. The most prominent appearance of the will (which I will uncapitalize for English speakers) appears perhaps in the work of Immanuel Kant.

Kant's idea is pretty straightforward. Unlike animals, which exhibit a kind of brutish drive, humans can guide their actions always and also for a wider range of reasons. The baker puts an extra bagel in your basket to make a baker's dozen. When she does so, she's fulfilling a norm of baking behavior. The driver takes the scenic route so that you might pass an elk. He's going out of his way to attend to your concerns. The wife slips her ring off her finger to give the impression that she is available. She signals her interest in outside affections. And so on. It's not that humans don't often aim primarily to satisfy their brutish desires; it's that human actions are sometimes also taken for other reasons. The will avails itself of some set of options constrained by extrapersonal reasons, just as we observed when we were shopping for dinner. Animals have no reason to balk at orphan meat. Humans do (or, at least, ought to).

The will is thus the steering capacity, the driving force behind a person's actions.[8] Whereas animals are driven by external heteronomous forces—instinct, desire, fear, sticks, carrots, etc.—humans are in a more important sense drivers of their own lives, subject to these external forces, of course, but ultimately autonomous (*auto* means "self"; *nomos* means "lawgiving"), capable of directing our lives *despite* these external forces.

Though the term *will* isn't in vogue anymore among philosophers, we use the word more often than we may realize. For one thing, it is an extremely common auxiliary verb. We use it to indicate what we *will* do: what we want to do, what we are disposed to do, what we're *willing* to do, what we're expected to do, and what we're supposed to do. We also sometimes use it synonymously with *shall*, suggesting that when we commit ourselves to doing something, that commitment is to be fulfilled in the future.

Our will therefore extends beyond us, our desires, and our earthly concerns. As such, it is no mistake that our *will* is also codified in an important testamentary document before we die, laid out as the full expression of our desires to be fulfilled long after we have passed on. Our loved ones and affiliates attend the reading of the will, when we impose our last bit of self-direction on the world of objects even though we are not around to want anything anymore. We will our children to go on for a good education, because even though they do not know it now, an education is good for them. We hope that our Aunt Betty will distribute our estate to those in need, because though we do not so much desire to help the needy while we are alive, we know that this is a worthwhile endeavor and should be done once we have passed and our earthly desires no longer interfere. The Kantian will, in essence, is in perpetual tension with the desires, subject *to* them but not directed *by* them.

Anyone who has ever gone on a diet can appreciate the tense struggle between willing and wanting. Those of us who love to eat but who also desire to lose weight must grapple with our will every time our bodies urge us to grab another cupcake. Fortunately, many of us exhibit at least a modicum of willpower. We may want to eat the cupcake, but we will ourselves not to.

We're actually pretty good at this sort of thing. We may lust after someone in our hearts, Jimmy Carter–style, but out of respect we deny ourselves the gratification of simply taking and using this person like a cupcake off the shelf. Under normal circumstances, it's not really a chore for us to practice self-restraint. It's part of how we respect people. Our animal impulses tug us in one direction; our humanity tugs us in another. The will, in this case, as in all cases, is the source of the *should*. The wants are there as a distraction from what we ought to do.

Being environmentally responsible sometimes involves a similar sort of struggle: a clash between determining our life trajectory (employing our steering capacity, our will) by appealing to the right (the justified, the most reasonable) principles and doing what is most comfortable for us, doing what comes naturally to us, and/or doing what we want. The trick to doing the right/justified/most reasonable environmental thing is (1) to figure out what sorts of principles, attitudes, values, rules, and commitments we ought to endorse, and (2) to get our wants in line with our wills so that it is no more a chore to be environmentally responsible than it is to respect other people.

Very often, students, politicians, and policy makers mistake the will as synonymous with desire. They think that if they will something, they just want that something more strongly than the extremely tasty objects right in front of their faces. So, for instance, they might characterize my decision to lose weight as a desire that's *stronger* than my desire to nosh on cupcakes. I am choosing this option, in other words, because I want to lose weight more strongly than I want to partake of fleeting happy-making deliciousness. Losing weight is simply a means to a different end: feeling good, appearing attractive to others, being healthy.

But this is a strange way of understanding the will, which we can see if we observe how frequently we simply want to want things that we don't really want.

Say what?

Yes. You read that correctly.

When I want to eat the cupcake but I want even more to lose weight, I want to not want the cupcake. This might even be true if I want to eat the cupcake but don't more strongly desire to lose weight—say my desire for the cupcake is so strong that I am driven to eat the cupcake *despite* my desire to lose weight. I might feel really terrible about indulging myself in a delicious treat that gives me a fleeting bit of happiness.

Consider:

I want to lose weight. That's true. I go to the gym, I lift weights, I jog approximately five miles every other day, and I try to eat healthy food. I do the minimum that I need to do to keep myself feeling and looking good. And, like most people, I do a lot of this grudgingly. As I pull on my gym shorts, strap on my shoes, don my sunglasses, and take off down the trail, I often think, "Just one hour from now, it'll be over." *One. Damned. Hour.*

I don't really like going to the gym or going for a run. Rather, I do it because I think I *should*. Living in Boulder, this makes me something of an anomaly. It's a running joke around here that almost everyone is training for a triathlon. We boast an oxygen bar, more climbing walls than you can count, and coffee shops uniquely marketed to appeal to bikers and long-distance runners. Just tonight I had to keep myself from asking the neighbor kids if they should be scaling the walls of their house. The situation is so bad that I have to plan my runs on secret trails so that I don't bump into any of my students or colleagues. I'm afraid I'd be ridiculed if they caught me thumping along like an injured Clydesdale. I marvel at the athletes around me who seem keen to step out of their houses to go for an eight-hour run. They love it, they say. They want more of it, they proclaim. If they could run all day and then go biking for the evening, stopping only once to refuel, they'd be happy.

Me? I hate it. I hate the exercise. I think they're crazy. I mean, sure, I'd *like* to like it, but I just can't bring myself to. It's painful and uncomfortable. I have better things to do.

In a certain respect what I really want is to *want* to run. I want control of this desire, and I struggle to get it. I go out for a run and tell myself that I can do it, that it's something that I like doing. I talk up my run with my friends, explaining that I prefer to run outdoors rather than indoors, on account of the change of scenery. I try to persuade myself of this. Usually I fail. Nevertheless, I keep going, hoping that someday I will overcome my dislike of the practice and get to the point at which I too can say that running gives me a great sense of satisfaction, that I love running.

Some philosophers characterize this as a second-order desire, a desire of a higher order, because it's not just that I want to run, but that I *want* to want to run. That's two orders of wanting there. There's what I want here and now. And then there's what I'd want if I were better. In a perfect world, my wants would be perfect wants, and I wouldn't have to struggle with them.[9]

There are many things that I might be said to want to want. I want to be a good environmental citizen, as I've said, so I want to want those things that will help me achieve this end. I want to want tofu, but it tastes pretty awful, so I'm in bad shape there. I also want to want celery, but I can't think of a less interesting vegetable. More difficult, I want to not want long showers, but I do prefer them, so I sometimes indulge myself. Many alcoholics, I suspect, want to stop wanting alcohol but are in the unfortunate position of being unable to do that. Sometimes we are fortunate enough to be able to bring our second-order desires in line with what we actually want. I want to like riding my bike to work, because it's healthy.

And, fortunately for me, I have both of these wants. I want to like riding my bike to work, and I do like riding my bike to work.

Here's another interesting thing about second-order desires. They're not usually the sort that limit their application specifically to me. I might also want my son to like broccoli, for instance, and I might want this for exactly the same reasons that I want to like it myself. Broccoli is healthy, and I want my son to develop a taste for things that are healthy. For the most part, he's a very good eater, so I'm pleased when I see him eating the broccoli that I've put on his plate.

So too with potentially bad stuff, like chocolate cupcakes and brandy. I really don't want my son to develop an affection for alcohol, now or in the future. Now, because he's a child, and in the future, because I'd prefer that he have a healthy relationship with booze, as everyone should. My concern about his future consumption of alcohol has quite a bit to do with my concern that he not become dependent on alcohol, that he not impair his capacity for self-control, that he not lose his unique human capacity to be a self-directed driver. If he develops an unhealthy relationship with chocolate cupcakes, this can also be bad. Too many of those ring-a-lings and he'll find himself stuck on the same ridiculous weight roller coaster that most of us know all too well.

The tension here between willing and wanting is the source of the *should*. Animals aren't subject to moral evaluation, to failing to fulfill an *ought*, because with very few exceptions, most of them act according to instinct and drive. But people are subject to moral evaluation, and the ever important distinction between actions and events is a reflection of this. Call this the "super-duper, double-secret, underlying principle of humanity."

The principle isn't really a secret, but it is deeply embedded in all of our actions and in all of our reasons for these actions, so it's easy to overlook. Whether those reasons refer to things we want and need basically or animalistically (and thus are brutish reasons) or whether they are reasons that extend beyond our wants and needs (and thus are normative reasons), they are always open to examination and articulation. Unlike animals, humans can presumably answer the question "Why did you do X?" Not only that,

THE SUPER-DUPER, DOUBLE-SECRET, UNDERLYING PRINCIPLE OF HUMANITY

People, unlike animals, act according to reasons; reasons that are, in principle, justifiable.

but it is an underlying presupposition of the distinction between actions and events that this reason is of the sort that it can be justified. I'll explain why this is important in chapter 9.

Still with me? Let's get back to our focus on being green.

YOU GOT YOUR NATURE IN MY PEANUT BUTTER

Advertisers never fail to remind us that environmentally complicated choices abound: Chicken or fish? Paper or plastic? Gas-guzzler or hybrid? We make choices all the time. But really, what are these choices? Some would have us believe that they are simply choices between preferences, as one might choose between a forest and a parking lot, between a river and a dam, between DDT and malaria. But I think there's something more to our choices. When we choose to do something, we are really asking wider questions than those concerned strictly with the consequences of our actions—or at least we should be. We are asking what reasons we have to decide one way or the other. Our arsenal of reasons for taking an action is our only entry point to a full ethical evaluation of whether a choice is right or wrong.

Return to the King Nut salmonella poisonings I mentioned earlier. For those raising their kids on organic vegetables and unprocessed foods, this can seem mighty confusing, if only in part because organic foods are generally thought to be healthy and safe. Parents who were providing their children with peanut butter weren't the victims of malice or hatred, and they certainly weren't the victims of a sociopath. They were victims of bad luck, of nature. It's hard to hold nature accountable, but there is an important sense in which we might want to hold someone accountable for peddling a bad product. In this case, King Nut, the US Department of Agriculture, or the Centers for Disease Control are the obvious ones to blame. And so we might be enticed into what the press likes to call a "blame game," pointing accusing fingers at a corporation or federal or state regulators.

This blaming exercise is not entirely pointless, however.

The decisions parents made to purchase organic peanut butter were only partly based on their belief that their kids desire and like peanut butter. They were obviously involved in many other kinds of assessments as well. Consider the various reasons that they might have had for purchasing organic peanut butter:

1. It's healthy.
2. It's more natural.
3. It's better for the earth.
4. It's tasty.
5. It's less risky.
6. It will win the parents the praise of their neighbors.
7. It comes in an attractive package.

I think most people will agree that some of these reasons are better than others. It's clearly good, desirable, right to keep kids healthy, but it's not right to do so simply because that is what everyone else is doing or because the packaging is attractive. It's also not clear that organic peanut butter is any healthier or tastier than nonorganic peanut butter. Some of these matters can be settled by observation and research: scientific deliberation or perhaps taste testing. But many matters will involve considerable conceptual engagement as well. Whether organic peanut butter is the "more natural" alternative is open to dispute.

The reasons on this list are unique because they aim at *justifying* the purchase of peanut butter. Thus they're *justificatory* reasons. They don't so much explain the behavior as offer a reason why we ought to engage in the behavior. They therefore stand in sharp contrast to the *motivational* reasons that investigators and detectives were seeking to extract in the Smith case. Sometimes the justificatory reasons line up nicely with a person's motivational reasons, just as when the will lines up with the wants; but sometimes they don't, as when people are tempted away from the reasons they might otherwise endorse, or when they act under mistaken ideas about what actions are justified. The important idea is that justificatory reasons run parallel to motivational reasons, but whereas motivational reasons *explain* behavior, justificatory reasons can only justify behavior. The great ambition of the ethicist is to find the right reasons, the justified reasons, and turn those reasons into the reasons that motivate us.

Just as choices are the junctures with which policy-making and decision theory is primarily concerned, justificatory reasons are the bread and butter of ethicists. Where many academic disciplines can explain to us how things work or what will happen if some lever is pulled, philosophical ethics has always been oriented toward asking questions about what we should do and how we should live—whether the reasons that guide us are justified. Whatever reasons we have for choosing one option over another, it is these justificatory reasons, not the motivations we harbor, that best explain our actions.

Many people, some philosophers included, would like to attribute these reasons exclusively to motivations and intentions, but it takes only a little reflection to see that the two are not necessarily linked. Only once we *endorse* and *adopt* such reasons do they become motivating for us, do they move us to action. More important, it is the reasons that we endorse and adopt, not the external causes that push our bodies around, that we evaluate when we evaluate a moral situation.

Randall Smith offers a case in point. The Appalachian Trail Murders were not simply a string of unfortunate events. They were a chain of choices—decisions, deliberations, reflections, endorsements, options—all mixed up with pressures created by internal and external influences. They were *actions*. If we really want to understand what happened in the Smith case, we not only have to explain what happened, we also have to be able to trace the reasons, rational or irrational, for what happened.

Correspondingly, to really understand how we should treat one another and what we should do for our environment, we would do well to approach the problem from the inside out, so to speak. We need to ask ourselves not how to model our decisions—as if we were looking down on our choices from an externalized God's-eye point of view—but rather what might distinguish some reasons from other reasons. We need a method of sorting out better reasons from worse reasons.

It is this observation about reasons—and the reasons-responsive capacity of individuals, the human will—that opens the door for us to drive a wedge between a simple approach to environmentalism, which assumes we have a duty to protect nature because it is so precious, and a more nuanced or down-to-earth approach, which takes into consideration the full range of reasons for protecting nature. This more robust picture of reasons colors every single choice we make: from decisions about our families to decisions about how to treat nature.

As I mentioned above, some ethicists think that we need to look only at the opportunities and outcomes of actions to decide which action is preferable. This view, *consequentialism*, has us plot out decision trees that compare better and worse consequences. But according to other ethical theorists, myself included, it isn't enough simply to aim at good outcomes. One must also aim at good objectives for the right reasons. This view, *deontology* (or duty-based ethics), has us evaluate a wider scope of reasons. We want to claim that there are morally relevant reasons that motivate our choices and thereby color their rightness or wrongness. We might have good objectives and good information, for instance, but still make morally problematic decisions. In general, the rightness of a decision should be evaluated according to criteria related to both the goodness of the objectives and the rightness of the reasons, a combination that will, of course, vary according to context.

CARNIVORE, OMNIVORE, HERBIVORE, LOCAVORE: PRINCIPLES AND HABITS

Strung out like popcorn on a fishing line, our choices can be plotted along a history of decisions. If we follow this popcorn string long enough, a trajectory starts to emerge: where we've been, and perhaps also where we're going. Far from a predestined path, this long string of popcorn is always under revision. Its trajectory is motivated not only by our immediate desires but also by much bigger questions: how we should live, what we should become, and what we should strive for. Our decisions about the best reasons are made in deference to these wider considerations and often in spite of our desires.

Why do some people become vegetarians, for instance? To reduce animal suffering, to reduce ecological devastation, to be trendy, and so on. In one respect, these are all goals and objectives, end states of the world. But it is a mistake to think that this is all they are. They are also reasons that underwrite a set of principles that guide action. Vegetarianism isn't accurately described as a simple set of eating behaviors, between those who choose salad over meat. It more accurately carves out a set of principled actions. Any old jackass can opt for a diet of salads, you might say, but it takes a little planning and ingenuity to be a vegetarian. To be a vegetarian, one must subscribe to vegetarian *reasons*. So long as it is these vegetarian reasons that are behind your eating behaviors, and so long as you are not mindlessly grazing on fruits and vegetables, it's probably fair to call you a vegetarian.

In contrast, it would bend the mind to refer to an animal as a "vegetarian" if it's the case that its diet consists only of vegetables. Just as animals can only ever be herded around by external masters, so too can they only ever be 'herbivores' (or 'omnivores' or 'carnivores' or whatever it is that best describes their diet). Animals don't make reasoned choices about what kind of lives they're going to lead. They're not self-determining creatures like us. The kinds of choices they make are choices only narrowly conceived, chasing their wants and desires. People, by contrast, choose in both senses: not only narrowly but also with the freedom to set the trajectory of their lives and to course-correct if need be.

When we choose to act for particular reasons, we sometimes implement overarching principles or constraints on our actions. Rather than simply choosing to eat this or that food, we place ourselves under specific dietary constraints: we'll be vegetarians, for instance, or we'll keep kosher. These principles—principles that we presumably don't just *have* but also endorse—are underwritten by a chain of reasons. If we choose to be vegetarians, we limit ourselves artificially away from our animal nature. We carve out a space for ourselves that is distinct from the rest of nature.

I want to eat a cupcake because cupcakes are delicious and because they make me happy (albeit in a fleeting sort of disposable way). That is why I *want* to eat the cupcake—for the experience; the cupcake is just the vehicle that gets me there. I do not want to eat the flank steaks of orphan children because it is not fair to the orphan children to offer up their flesh for consumption. This is the reason I do not want to eat orphan children; it stems from my endorsement of a principle. It would be deeply selfish to consider only my interests in eating a single meal of orphan meat when there are so many reasonable objections that an orphan may have against my decision to eat him or her, or that someone else may raise on his or her behalf.

For this reason, I am (and presumably we all are) non-orphanitarians. We subscribe to the principle of non-orphanitarianism because there are a million reasons that eating orphans is wrong. We gain control of our bad habits by forcing our behavior away from the animalistic desire to eat whatever lies before us.

Reasoned choices don't stop at the dinner table, of course. They are part of our workaday life, part of our family life, part of our sociopolitical life. They appear at an individual level (with the things that you and I each decide to do) and at a social level (with the things that we all decide to do collectively). Just as you or I decide what to do, so too do larger institutions decide what actions to take and what laws will guide them. These little choices I was talking about above, these everyday decisions, are the bread and butter of every program administrator and policy wonk; and in the formation of public policy, the policy wonk aims to design policies that steer individual decisions.

Consider now how the will has manifested in political theory. Jean-Jacques Rousseau noted many centuries ago that we can speak intelligently of a *general will* to describe the overarching perspective of the collective. This general will is often interpreted as synonymous with a *general welfare*, or some such related notion, but this is incorrect. Rousseau's general will is much more abstract, much less rooted in the desires of the majority, and much more aimed at pinning down what the general citizenry ought to do if it were to act as one body. According to Rousseau the general will's most natural expression is in codified law. If correctly written, laws and policies reflect the general will and steer the social body in much the same way that the individual will steers you. When countries establish environmental laws, like the National Environmental Policy Act or the Endangered Species Act, they put into motion rules that guide their decision making, much as you institute a rule that guides your decision making about cupcakes. In principle these laws are supposed to keep us on the straight and narrow, to help us avoid the enticements of cookies and candy and of economic benefits that so often muddy the successful implementation of good public policy.

No doubt, the will—the private will and the general will—is undoubtedly a controversial notion, but it (or its modern-day variant) is vital to understanding ethical problems. It is, essentially, the *bearer of reasons*, standing apart from the simple goal-oriented drives characteristic of animals and instead functioning to provide rational support for the things we do. According to many ways of thinking, it is only people, and only collectives of people, who maintain the capacity to express their wills in the formation of law.

The question about what we ought to do, regarding the environment or anything else, can be approached best by first reexamining our conception of external factors (states of the world, states of our environment) and understanding that they can be just as physically internal (states of mind, drives) as physically external. In other words, the external factors that push a person to do X, Y, or Z are external only insofar as they are not sourced from within, from the will. An addict, for instance, can be driven just as much by his need to get his next fix as a prisoner may be driven to work the fields or, perhaps, a wildebeest may be driven to sate his most primal needs.

By contrast, if we turn our attention more directly to focus on reasons, and specifically acknowledge that we are humans who can share reasons, justifying our actions in the process, I think we can maintain our grip on our humanity while also acknowledging that it is out of our humanity that our commitment to a clean and healthy environment must emerge. Placing constraints on our feckless desires and guiding our actions by appeal to reasons isn't at all an abandonment of who we are; rather, it is a further assertion of our self-legislative capacity, of our will, of our autonomy. In fact, our own capacity to endorse some principles and reject others, thus shaping an identity around our commitments—to cut back on our emissions, to eat more veggies, to be nicer to our neighbors, to be better parents, etc.—may very well be what keeps us human, what defines us as free.

It has always struck me as curious, then, that so many environmentalists seek to bring the natural and the human back into harmony, as if to deny that there's any important distinction between animals and humans, as if to deny the reason-bearing will.

Most people want control over their lives, I assume. They want control over what they do and what happens to them. They don't want to relinquish more control or be herded about, either taking their marching orders from elsewhere or having their actions interpreted in this way by others. That so many branches of environmentalism encourage humans to accept and embrace their animal nature, to the exclusion of their humanity, may be partly what rubs so many people the wrong way when they are told that they should get back to nature. It's not that nature isn't worth getting back to, or even getting in touch with; it's that environmentalism of the "precious vase" variety appears to urge us to relinquish our humanity and our autonomy.

6

Dr. Feelgood and Mr. Fix-It Go to the Picture Show

In which the author (1) introduces the specter of target-based environmentalism, (2) discusses the sundry broken promises of utopia, (3) offers a criticism of the view that our problem with aiming at utopia amounts to unwarranted risk taking, (4) proposes that the good is the handmaiden of the perfect, (5) introduces an alternative explanation for those who resist vaccination, and (6) suggests that this alternative explanation can help us gain a grip on justifications for a variety of environmental interventions.

THE BIGGEST FATTEST AWESOMEST AIR CONDITIONER MONEY CAN BUY

Whatever it was that inspired Kurt Vonnegut's older brother, Bernard, to shoot rockets filled with silver iodide into cumulus clouds in order to compel rainfall over thirsty farmland, it was likely nowhere near as menacing as the threat of anthropogenic climate change. Bernard's research was probably motivated by considerably more mundane concerns: a simple fascination with the inner workings of weather systems, the promise of extraordinary profit, or a megalomaniacal urge to dominate the universe. I doubt the latter, but it's hard to dismiss it as a possibility.[1]

The young Bernard could little have imagined that his early experiments in cloud seeding were laying the groundwork for a far more ambitious project that would not only affect the weather but also radically reshape the earth's climate. Over the past several years scientists and policy makers have gathered to discuss a suite of options and technical solutions to the climate problem. Their thought is that we can geoengineer the climate back to global mean temperatures that line up better with the historical record. The proposals for such technologies vary, but they range from simple low-cost solutions, like dumping tankers of iron filings into the open sea (ocean fertilization) or injecting sulfur dioxide into the troposphere, to much more expensive and science-fiction solutions, like suspending giant mirrors above the earth to reflect sunlight. Technological visionaries spin tales of an easy climate fix, enticing some but upsetting many others. Critics worry that ocean fertilization schemes will destroy life in the oceans, or that tropospheric sulfur dioxide injection will authorize continued pollution and generate torrents of acid rain.

There are many compelling scientific and technical arguments both for and against geoengineering, which is probably why discussions of it continue in earnest. But even if our best science indicates that geoengineering can succeed technically, there are clear ethical reasons to rule it out, since it can never meet with the scrutiny that most of us take to be emblematic of justified right action. The problem with most of these technical arguments, in my opinion, is that they focus so strongly on the end-state of the world, on the world they will bring about, and ignore, for the most part, considerations regarding the responsibilities of individuals or collectives to limit their actions to those that respect the rights and concerns of others, now and in the future.

Consider, for instance, one of the more prominent proposals: ocean fertilization. First advanced in the 1980s by Woods Hole oceanographer John Martin, the thought is that we can dump several tankers of iron filings into the sea in order to manufacture a midocean algae bloom. Researchers project that such an algae bloom might then suck carbon out of the atmosphere, much like a ShamWow sucks soda from the moldy underbelly of your basement carpet. All of this sounds mighty enticing when you consider the unpleasant climatological upheaval that is slowly unfolding and that will fundamentally change the world in which our children live. Yet given the complexity of ocean ecosystems and humanity's reasonably embarrassing failure rate with ambitious engineering projects—the Panama Canal mosquito eradication project, the Everglades restoration project, the Project Stormfury attempt to weaken tropical cyclones by seeding them with silver iodide, to name just a few such failures—there's plenty of reason to worry that tinkering with nature in this way may be ill-advised.[2]

Rainmakers like Vonnegut were mere redistributive Robin Hoods, stealing rain from the rich and giving it to the poor. But the latter-day heirs to such research propose no simple hydrological deck-chair shuffling. They aim to fix one mess not by straightforwardly cleaning it up but by introducing another mess. They aim to pollute our oceans, in other words, in order to clean up air pollution. In doing so, they threaten to either sink or save our ship. Fertilizing the oceans runs a real risk that the citizens of this planet could fall victim to the same fate that eventually nailed the old lady who swallowed a fly: we could get caught up in an endless chain of curatives, repairing one problem only to introduce another. But that's only if we blow it. If we get the science right, we could break the chain. We might have at our fingertips a relatively cheap way of reversing the atmospheric concentrations of carbon that the past hundred years of industrialized recklessness have left hanging over our heads.

In the previous chapter I spelled out the ways in which individual actors can deploy their wills to exert control over their wants and desires, identifying this as the signature of actions—evaluable, unlike events, from the standpoint of morality. I also suggested that a similar principle-guidedness can be exerted over actions, policies, and interventions at the local level, that all actions are therefore best assessed not in terms of their outcomes or ends but in terms of the principles that underwrite them.

But I want to be clear. When I suggest that we need to look past our wants and desires, the case I'm building isn't ultimately about control. It's more about ensuring that we act for the *right reasons*. And herein lies our next major confusion. We tend to think that in order to overcome our desires we should set concrete rational goals for ourselves: 1,200 calories per day, five pounds by the end of summer, two sizes before New Year's. Indeed, it is this simpler picture of choice that has informed so many in the environmental movement. They've visualized a better future or an ideal world—ecotopia, Walden Two, sustainable development, 350 ppm (of pollutants), 2°C, zero population growth, carbon neutrality, etc.—and directed their policies to hit this target.[3]

Overcoming our environmental problems, on these views, is merely a matter of identifying goals, acquiring more and better information, educating the voting public about projected outcomes, and changing widely held beliefs about the impacts of our actions. Obviously this is no simple task, and we spend unfathomable amounts of money and energy trying to meet this lofty ideal. But what if this approach to environmentalism is totally misguided? I think it is.

So where earlier I was concerned with intentional action and the moral ambiguity of desires, here I want to focus on the moral vacuity of beliefs, of knowledge, and even of facts about the best states of the world. Specifically, I want to focus on the technocratic side of environmentalism, on those who think that we can resolve our environmental issues through a technical fix; on those who think that maybe we ought to shoot for identifiable objectives—a happier, better, healthier world, say—but who do not realize that these objectives are thrown off course by false beliefs, uncertainty, and/or risk.

THE ELIXIR OF UTOPIA

The Death Star, I suspect, is a pretty unpleasant place to raise a child. Notwithstanding concerns that Lord Vader may trample the family puppy, there's the clinging and clanging of pipes, the infernal echo, the creepy breathing, the errant laser blasts, and rodentlike robots skittering underfoot. Yet despite these trifling annoyances, the Death Star is a perfectly contained, perfectly functional ecosystem—the dream of urban planners everywhere.

In the mid-1960s, roughly about the time of Rachel Carson's much ballyhooed fire-ant and songbird cataclysm, an American visionary began charting plans for a model community, a city of tomorrow, that would reinterpret the mistakes of the modern American suburb to build a hybrid city-town-supervillage—a habitable paradise in an otherwise barren and lifeless parking lot. The blueprint for this planned community was to be informed by the successes of previous cities, but it was to be left open-ended so that designers could continue the hunt for new and alternative ways to live communally. It would be an ongoing experiment, a perpetual unfinished project, with shops and homes and streets built up and torn down so that urban planners could unlock the secrets to life in the future, "introducing and testing and demonstrating new materials and systems."

Joy. I don't know about you, but even just a few months of road construction in my neighborhood makes me ready to put my house on the market.

Whatever the case, the idea was that this model community would draw from the creative energy of industry and enterprise in an evolutionary manner, keeping what was good and discarding what was not good, adopting John Stuart Mill's maxim that all good ideas float to the surface whereas all bad ones sink to the bottom. As such, this evolutionary vision was in essence a revolutionary vision, not too mired in any particular picture of the ideal city, but an attempt, nevertheless, to make utopia a reality.

Unlike most dream cities, however, the experimental prototype community of tomorrow wasn't left as a mere idea. Two months after its forward-looking creator had passed away, dozens of well-intentioned collaborators, colleagues, and companions dusted off his notes and, in homage to the great visionary, proceeded to screw up the futurist's project so colossally that had the as-yet unnamed designer been cryogenically frozen, as gossips and gadflies would later allege, his boiling blood would surely have thawed his icy corpse immediately.[4] The botched realization of this ideal neighborhood opened its doors in a location not very far from the beloved stomping ground of little children, exhausted parents, and ten-foot mice. Apart from the enormous geodesic buckyball that is its inexplicable signature, most of us have few other idealistic associations with what we now call Epcot.

Vader and Mickey are infrequently mentioned in the same context, but they do share an affection for self-contained ecosystems as well as giant floating buckyballs. The freakish globe crowning the main plaza in Epcot has its origins in Walt's visions of placing his entire experimental city under a dome. It's not clear how the buckyball—relatively small, compared to the grounds of Epcot, and extraordinarily small, compared to the Death Star ("That's no moon; that's a space station")—is somehow supposed to make up for the fully domed city, except that Buckminster Fuller, alongside Walt Disney and Darth Vader, may have been the one serious scientist to advance this idea.

Indeed, at about the same time that Grandpa Walt was drawing up plans for Epcot, Fuller was busy trying to persuade urban officials to build a fantastically large cupola over the city of Manhattan. Consider the possibilities. New Yorkers could have perfect climate control. In the middle of a summer afternoon, no one would collapse from heat exhaustion. In the spring, hawkers of cheap $3 one-use umbrellas would go out of business. During winter snowstorms, no shoes would get wet, no hair would get mussed, and no cabs would splatter innocent bystanders with a wave of slush and street runoff that collects around the curbs as the ice melts. Brilliant!

Were this the only half-baked scheme to hit the drawing board, I suppose it might stop there. But dreamers and visionaries have for centuries been imagining and designing improvements on the city, on our built environment, almost all of which involve improving human welfare by keeping nature out. In 1979, planners in Winooski, Vermont, proposed that if a dome were built over the whole city, the town could save 90 percent on its heating bills.[5] Silly as it sounds, some have even proposed that a giant dome be built over Houston. Beyond even the dreamer phase, the O2 Millennium Dome in downtown London boasts a diameter of 365 meters (400 yards) and covers almost 25 acres of land (which is about six city blocks).[6] Its name, the O2, clearly evokes control of oxygen and the atmosphere.

The push to build utopia emerges from a need to solve problems that are primarily synonymous with diminished well-being. Consequently, many of these urban utopias presume that such problems have a technical fix, which in a certain respect makes sense.[7] If you're talking about the mechanics of the universe—snowstorms, earthquakes, floods, hurricanes, beasts, malaria, waterborne pathogens, climate change—a mechanical solution may be the only way forward. To solve the problems presented by nature, build bigger levees, stronger walls, and taller fences.

Historically speaking, a utopia is a place of great good, where all or most are happy. Thomas More is widely attributed with having written the first novel of this genre. His famous work from 1515 actually coined the term. The word *utopia* comes from the Greek words *eutopos*, meaning "good place," and *outopos*, meaning "no place." In common parlance, it has come to refer to a paradise or an arcadia.

Not long after More's work, Sir Francis Bacon, the great-granddaddy of the scientific method, described another kind of utopia in *New Atlantis* (1624), offering a view of an intellectually vibrant world structured around discovery, generosity, enlightenment, dignity, piety, and public spirit. Of course, many politicians and political theorists before and after these foundational figures built their careers on dreams of a utopia. Socrates imagined a just republic structured in part around a benevolent lie. Marx wrote of a worker's paradise. Vladimir Lenin wrote in 1912 an essay called "Two Utopias" in which he argues that most utopias are mere fantasies. "Daydreaming," he says, "is the lot of the weak."[8]

With few exceptions, the various instantiations of utopia have focused primarily not on accepting and embracing nature but on crowding nature out, on cleansing nature of the bad by picturing a more pristine, better environment that is an improvement on the natural environment. Just as elevators and air conditioners have improved our well-being in some respects and degraded it in others, the same sort of appeals to a "better world" drive a majority of environmental policy arguments. Whether such arguments are rooted in the promotion of the general welfare, the optimization of overall economic benefits, or even the conservation of nature, starry-eyed dreamers specify the place toward which our resources and technologies should be directed and leave scientists, engineers, and practitioners to step in and sort out the details. This is, after all, how many interventionist regulations and policies are defended in welfare capitalism: through the suggestion that the general welfare can be promoted by tweaking markets so that they do not fail to achieve the most optimal outcome.

HENNY PENNY AND THE UGLY DUCKLING

As a consequence of this common presumption that morally right actions and policies can be justified in welfare terms—a presumption which drives many policy initiatives: "If we just pursue such and such a policy, the world will be so much better."—it is common for critics of engineering projects to argue against these projects in terms of risks and unintended consequences. In doing so, they imply that if the risks could be minimized, the projects might then be justified. Many people, for instance, will protest that the biggest problem with geoengineering is that there are still too many things we don't know about the climate, the earth, and the engineering. "The risks are too great," they will say. "There may be unintended and costly consequences." This has been the predominant thrust of the geoengineering discussion since its beginning.

In my opinion, this line of reasoning rests on a mistake: the belief that our environmental issues are best understood in terms of welfare.

Consider, for starters, how slippery the term *pollution* is. Most people instinctively define it as emissions that cause harm or damage. But it depends on your perspective whether all such emissions should be considered pollution in the first place. There are lots of different kinds of emissions—point source, area source, mobile source, and natural source—and some of them confer benefits. To most farmers, for instance, the addition of organic compounds like nitrogen to the soil would be a gift from the gods, dramatically improving crop growth and foliage. Too many of these compounds however, and uh-oh, the crops die. In one case they're essential; in the other, they're a pollutant.

The same can be said (and has been said!) for carbon. As many foes of environmental legislation disingenuously harped when the Environmental Protection Agency (EPA) proposed in 2009 to regulate carbon dioxide like a pollutant, carbon dioxide is essential to life.[9] It helps plants grow. Their claim was that carbon dioxide cannot be a pollutant because it is not always harmful. What they were doing, however, was taking advantage of this flimsy definition of *pollution*. The reason for this terminological slipperiness is that pollution is typically framed in terms of harms and benefits, making its categorization entirely contingent on whether the affected party will be made better or worse off. The economist Ronald Coase, in somewhat more sophisticated fashion, pointed this out decades ago. Carbon dioxide is the same way: essential to plant life, but when unmitigated concentrations of it invade the atmosphere, it has the undesired effects that we are witnessing now.

Problem is, we don't always know how welcome these harms and/or benefits will be. One farmer may need more of one compound for a certain project, and another may need less. It is presumptuous and morally suspect to make assumptions about what is right for them. Moreover, it is flat wrong to assume that a particular action is ethically permissible just because it may confer overall benefits, as advocated by, say, Mill's "Greatest Happiness Principle," which proposes that actions are right when they promote the greatest happiness of the greatest number affected.

Consider the following: If I wake up from knee surgery to a smiling surgeon who enthusiastically informs me, "While you were asleep, we went ahead and added a pacemaker to your heart, just to be on the safe side," I might have great reason to feel wronged, even if the pacemaker is 100 percent safe and even if I am physically better off. Or consider this: If I return from vacation to learn that my neighbors, who are college students, have repaired the walls and furniture in my house, perhaps after they and 100 friends of theirs have had a raucous party during which overflow partygoers had broken into my house and damaged my property, I may again feel wronged. Perhaps my neighbors have made me better off than I was before, maybe even by making improvements to my property. One would think I'd be grateful for the free labor. But there is a strong sense in which I would believe that they had only heaped one wrong on top of another.

What makes an action right is not just whether that action makes the world better, but also whether those who will be affected can or could agree to having their world made better by others. If my house was trashed because of my neighbors' party, perhaps there are other remedies that I would like to explore that would be more appropriate for me, my family, and my property. If my neighbors take the initiative to repair my belongings without consulting me, they usurp my control over these possibilities, and in doing so they disrespect me and violate my right to do otherwise. They suddenly bear the responsibility for having changed something in my house that might have been reversible in another way that was more to my liking.

Just so with many geoengineering technologies. Even though ocean fertilization might in fact make the world better, we need to ensure that the people who will be affected by these improvements could all agree to them. If a giant algae bloom generates enough food to spark wonderfully delicious and nutritious new fisheries, that may be very good for the world, insofar as it yields extraordinary benefits; but there are still strong rights- and respect-related ethical objections to aquaforming our oceans in this way.

Focusing on risk and unintended consequences completely brushes aside these other important ethical questions. It shifts the burden of proof for what we ought to do onto the shoulders of the engineers, scientists, and fortune-tellers, reducing the social issues associated with geoengineering to primarily a technical matter: whether it can be successful.

Nassim Nicholas Taleb's excellent book *The Black Swan* covers the phenomenon of uncertainty at length, suggesting that almost all revolutionary changes in the history of modern civilization have come about unexpectedly. Taleb is correct in proposing that there is much we cannot know for certain and in suggesting that those who emphasize risk place an incredible burden on knowledge; but one of the great advantages of the scientific method is that it empowers us with at least a modicum of predictive capacity. If I throw a rock at a glass window at a certain velocity, I can be pretty sure that the window will break. If I overload a bridge, I can anticipate with some certainty that it will collapse. If I see that a dark storm cloud hangs portentously over Tuscaloosa, I can feel confident that it will rain. Existing uncertainties are often just a call to do more research. In most cases, all it takes to overcome the uncertainty, and perhaps to minimize or remove the risks, is sufficient research. Many people thus take these scientific certainties as worthwhile gambles.

Sure, risk is a major concern with the geoengineering technologies, just as it is a concern with any new invention. We should all be worried about the implications of our actions, about the risks of destroying, or at least dramatically altering, the oceans and the climate. If ocean fertilization will create a scenario in which the oceans become uninhabitable for most fish and aquatic wildlife, this is clearly an unacceptable outcome, and we ought not to proceed. But the science is unclear on this outcome, and there is strong evidence to suggest that we can fertilize the oceans without making a mess of things.

It is my view, however, that *even if* ocean fertilization were to yield a far more palatable outcome—perhaps by producing enough algae to generate a banner fish harvest, thereby not only reversing climate change but also feeding the world's hungry—there are still strong ethical reasons not to use it as a method for reducing greenhouse gas pollution. These reasons include not only damage or benefits to specific populations but also encroachments on the lives and livelihoods of others who may not welcome those encroachments. Very much like my neighbor who improves my property without consulting me, we can be wronged even though we have not been harmed and might even have benefited.

The point here is twofold: first, that environmental uncertainty and risk, while legitimate concerns, do not in and of themselves militate against a proposed environmental intervention; and second, that appeals to risk and uncertainty cede too much moral ground to scientists and engineers; folks who happen to be very good at knowing and fixing things but are not necessarily sensitive to the wide spectrum of rights and responsibilities that might be neglected or trampled during the process of knowing and fixing.

WHY HONEY, IT'S PERFECT!

The original inspiration behind Epcot was the much grander vision of reinventing the American subdivision that I described above. It was one of many attempts at a new urban lifestyle, known as New Urbanism, unchained from the unpleasant discomforts of most cities. The living center of Epcot was to be based within a small town, and the small town was to be equipped with commercial and residential spaces, hidden driveways, and easy public transportation.

For a small and select few, Walt Disney's dream was eventually realized in the form of a different model community and christened with a somewhat more saccharine name: Celebration. It is located about five miles south of Epcot, two exits down Interstate 4.[10]

Celebration is, in many respects, an extension of Disney World. Depending on your attitude about amusement parks, this could be a wonderful thing or an abomination. Me? I'm thinking it's not so great. In principle I might not mind huffing my amusement from a paper bag, but I suspect that if I actually lived in the plastic cookie-cutterville that is Disney's Celebration, I'd go bonkers. I'm not positive about that, because apparently the people who live there do love it. But then, people love all sorts of wacky things, and dare I say that many of the things that normal people love—fast food, fast cars, shiny objects—leave me cold.

It's just so damned pristine, replete with perfect parks, perfect porches, perfect lawns, perfect ponds; everything is perfect.[11] In the fall, the planners of Celebration even simulate the perfect falling of leaves by blowing scraps of colored paper into the air, since the Florida climate doesn't allow for broadleaf deciduous trees. It's a masterminded, overplanned community. There's nothing natural about it. This shutting out of nature is the most disconcerting thing about Celebration.

One may be inclined to believe that such imperfection in perfection is simply a problem of execution: the engineers and builders have somehow blown it, they haven't gotten something right. If they could just get better-looking leaves, if the colors of the mailboxes only matched the rain gutters, if the weather were just slightly more uniform, then all would be good. But every utopia suffers some problem of execution; even the most perfect circles, when drawn carefully on paper, exhibit imperfections. I suspect it's not so much a problem of execution as a problem of conception, which has been illustrated, time and again, through the philosophical magic of the literary dystopia.

Novels that are considerably more pessimistic about utopia, in which positive aspirations have driven well-executed and administered utopias, tend toward dystopias: negative utopias, often premised on the very principles that underwrite the positive utopias that are the subject of so much dreaming. Swift's *Gulliver's Travels,* Huxley's *Brave New World*, Orwell's *1984* and *Animal Farm*, Zamyatin's *We*, Atwood's *The Handmaid's Tale*, Burgess's *A Clockwork Orange*, Bradbury's *Fahrenheit 451*, Wells's *When the Sleeper Wakes*. Not all of these are instances of worlds that were too risky or that yielded unintended consequences, but they're almost all cases in which some value was promoted to the exclusion of other values and principles.

Huxley's *Brave New World* is a utopia built around happiness, and it's fairly successful at painting a picture of how such a utopia might be constructed. The grand irony is that this utopia, though built specifically with the objective of promoting happiness, is almost entirely devoid of happiness. Although it succeeds at creating an environment in which humans have no desires that go unsatisfied, the inauthenticity of these desires leads the reader to question the utility of their fulfillment. Atwood's *Handmaid's Tale* is another utopia built around a particular view of Christian virtue. It too is fairly successful. Again, however, it's just a terribly vicious place to be, where laws are enforced by the boot heels of bastards. Orwell's *1984* is a utopia built around security and safety, allegedly in the name of preserving freedom, yet it is the quintessentially unfree society. In almost all cases, dystopias depict utopias that have gone awry, but not because the outcomes were unpredictable or there was some risk that the utopia might turn out differently from what the planners had expected. Rather, they have become dystopic because they were built around one conception of the good. In many cases, inherent to this conception of the good is some internal contradiction.

Voltaire is commonly quoted as saying that "the perfect is the enemy of the good." But I think differently. I think that the perfect is inextricably intertwined with the good and that you can't have one without the other. The good, essentially, is the handmaiden of the perfect. Once you introduce some conception of the good and drive headlong toward the attainment of that good, if you ignore, as you drive, the justifiability of the actions that you must take to achieve that good, you unwittingly override the vast landscape of obligations and commitments that otherwise stand like signposts and barriers along the moral road. In essence, you crowd out vital considerations related to constraints on actions and responsibilities to others.

What I mean by this is that a relentless pursuit of the good can just as easily override considerations of the right—questions about permission, consent, duties, obligations, and rights—as it can bring us to a state of the world that is shiny and perfect. One way to better see how the right can be overriden is to push the conceptions of the good to their perfect limits, to the point at which dreams of utopia yield to the realities of dystopia. If we do this, as Huxley did, we can see more easily that an outcome that may at first seem entirely good and acceptable would probably, when taken to the level of perfection, be deemed unacceptable by many people who would benefit from it. The test for justifiability should be whether the world we are aiming at would be acceptable to all.

Despite my doubts about achieving a utopia, I have high hopes for New Urbanism. In principle, I think it's exciting and outstandingly cool. I desperately want it to work, and I privately harbor a dream of living in a New Urbanist community someday. Most of the things I love about big cities are incorporated into the architecture and planning of the otherwise small and futuristic towns. And most of the things I love about small-town life are there as well. New Urbanism is a big-idea movement built around the implementation of the best of both worlds. Take the best of one world, combine it with the best of another world, engineer the two so that they work in harmony, and, presto! Call it a community.

My claim here isn't so pedestrian or speculative as to suggest that Walt Disney or Buckminster Fuller or the various other utopians were in fact motivated by a sense of the good. We simply can't say what motivated any of these futurists. We can only report what they tell us. We can retell the story that they've told others. But it is just as plausible that they were motivated by a desire to earn a lot of money or to become famous or to prove to their parents that they could make it as visionaries. For the same reason that we cannot say for sure what motivated Lord Vader and his behelmeted henchmen but can tell a compelling story about their conception of the good—which in that case was basically the inversion of the good, though it was still a "conception of the good"—we cannot say what motivated those whom we otherwise hold in high regard.

To see more clearly what I'm saying, consider how complicated this reasoning can get by looking briefly at another domain closely associated with welfare: health.

FLU SHOT ROULETTE

Pediatricians of the modern variety can rattle off a list of utterly terrible childhood diseases from which all parents would do well to protect their kids: measles, mumps, rubella, hepatitis, rotavirus, diphtheria, tetanus, pertussis, pneumococcus, meningococcus, polio, varicella, influenza, and so on. These are, more or less, well-known and well-understood diseases. Fortunately, science has given the modern pediatrician a suite of vaccines with which to address these afflictions: MMR, HepB, DTap, PCV, IPV, and so on. Almost all of these vaccines are incredibly safe, and in all cases of the aforementioned disease/vaccine pairs, the American Medical Association (AMA) and the American Academy of Pediatrics (AAP) recommend immunization of children under six, with various age-appropriate caveats.

Like many medical interventions, these immunization recommendations have their proponents and their opponents. In my neck of the woods it is not uncommon to hear parents grousing about vaccinations as they sit on park benches feeding their tightly swaddled children unpasteurized goat milk.[12] Not the smartest move, all told, but hey, what's a little E. coli and Campylobacter between a mother and her child?

One particularly contentious vaccine is the influenza vaccine, said by some to cause this or that ailment; the afflictions vary. Despite the ever tolerant public sphere, these vaccination pessimists are ridiculed by well-meaning scientific types for being irrational or selfish, and they are often taken to task for not getting the math right. "The odds of a negative outcome from the flu," many say, "are much greater than the alleged downside of the vaccine." But I think that the vaccination pessimists aren't as crazy as they're sometimes made out to be, so I'd like to argue here on their behalf.

(I personally don't have suspicions about most vaccinations, but a non-insignificant number of intelligent and well-educated people do. Understanding how this discussion derails will, I hope, shed critical light on some misunderstandings that are currently in play in the environmental movement.)

Maybe I'm nervous that I will harm my child if I give him a vaccine, for instance, and I don't want to be responsible for harming my child in this way. "Are you sure that the vaccines are safe?" I might ask my pediatrician. I can have lots of information and the right values but still have a reason to doubt that I should get my child vaccinated. If this is the problem for the vaccination pessimists, then it's not simply matter of collecting more and better information, dispelling uncertainty, and overcoming issues of trust, as many scientifically inclined vaccination optimists are wont to assume. It's not that these parents don't understand the risks or the harm—it's that they don't want to be *responsible* for harming their child.

If this is the problem for me, it doesn't matter that most of the scientific studies point in the direction of medical safety. What matters is whether I'm going to be safe from blame. Only in some cases is a more sophisticated scientific discussion going to get me to believe that vaccines are safe. Sure, they may indeed be medically safe, but if I screw up or I'm unlucky, are you going to carry my guilt?

Maybe this point is a little difficult to see. I have a fun case.

Suppose the dread schmoo, a disease so vile and nasty that it will kill a child within 24 hours of contact. The schmoo, not unlike many diseases, is a mutant variation on earlier vile plagues, and has arrived on domestic soil via the muddied shoes of some wandering spirit who, during a reckless weekend sojourn, traipsed through the jungle of Nool and refused all in his homeland the plain courtesy of wiping his feet on the welcome mat. Like many diseases, schmoo is spread through standard airborne means. One breath from a vulnerable human and the end is nigh.

The schmoo, bad as it is, is relatively uncommon. Any given person stands an epidemiologically nonnegligible chance, say 1 in 10,000, of contracting and dying from the schmoo, fortunate as we are to be the beneficiaries of soap and restrooms and now ubiquitous hand-sanitizer outposts.

Fortunately for humans everywhere, intrepid researchers have developed a successful vaccine that will alleviate all symptoms and all ill effects of the schmoo in all potential victims. Huzzah! The vaccine is a success, but it carries with it a substantial downside. Where on the upside it will completely alleviate all cases of schmoo, there is an epidemiologically nonnegligible chance, say 1 in 20,000, of suffering an allergic reaction to the schmoo shot and dying from the reaction. In effect, the administration of the schmoo shot cuts the death toll in half, which is a mighty fine policy outcome.

Vaccinate the entire population?

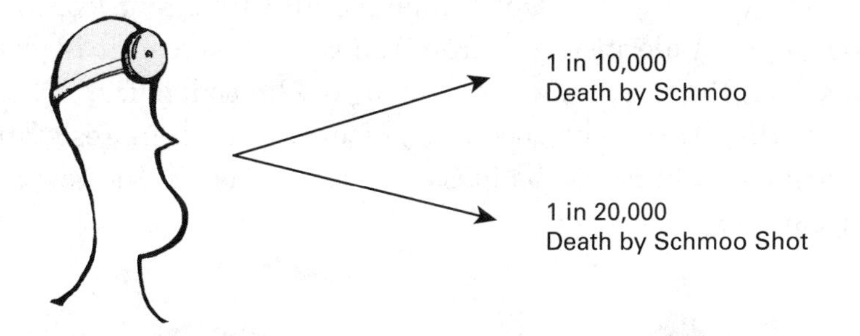

Whereas any individual stands a 1 in 10,000 chance of contracting and dying from schmoo, that same individual stands a much more desirable 1 in 20,000 chance of having an allergic reaction to and dying from a schmoo shot. It would appear to most people that this is a no-brainer gamble. We should all take the chance on the schmoo shot. This makes sense. The outcome is twice as bad for abstainers from the shot as for those who take the shot.

This is the line of reasoning that supports widespread vaccination programs, some pesticide-spraying programs, and almost any environmental intervention you care to name: the construction of levees, the cutting of firebreaks, the building of retaining berms, and even the geoengineering of the climate. The idea, clearly, is that we can alleviate some environmental risk by intervening, and in order to determine whether we should intervene, we need only look at the costs, benefits, and risks.

But I think the answer to the puzzle isn't really entirely clear.

From an aggregate standpoint, it is obviously worse to have twice as many deaths without the schmoo shot as with it. From an individual perspective, the odds don't tell the whole story. Consider the following more concrete variation.

Nature's card dealers, ever the cynical sort, have dealt a confounding doozy of a hand. Unbeknownst to researchers, the cause of death by schmoo or schmoo shot is the result of an obscure and hitherto undiscovered chromosomal variation. All carriers of *goo*, one of these undetectable chromosomal variations, will die of schmoo if they come into contact with it. Evolutionarily speaking, goo carriers have not yet developed resistance to schmoo as most members of the general population have. A simple schmoo shot can counteract their goo vulnerability. Unfortunately, all carriers of *foo*, another variation, will die instantly from the schmoo shot if it is given to them, even though they are resistant to schmoo itself.

In other words, the goodness of fooness lies in its resistance to schmooness, where the badness of gooness lies in its newness to schmooness.

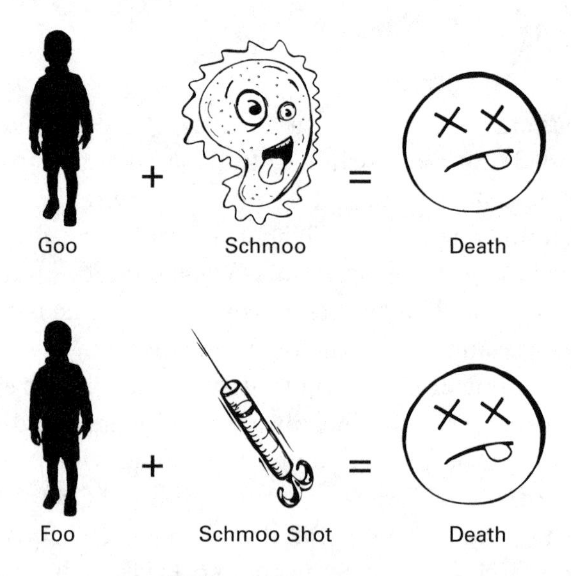

The allergic reaction to the schmoo shot manifests only in those who are not vulnerable to the dread schmoo itself, since their immune systems are strong enough to fend off the invasive virus, but alas too weak to handle the overload of antibodies. Had they no shot, they would be safe. There is no way of telling whether one will contract schmoo or have an allergic reaction to the schmoo shot.

Knowing this information, but having no information about whether you are a goo or a foo carrier, what should you do? I ask not what you would do as a public health practitioner, but as an individual. Would you, or would you not, take the shot?

If you answered that you *wouldn't* take the shot, I would be curious as to your reasoning. I suspect it might have something to do with the odds.

If you answered that you *would* take the shot, what odds might bring you to *resist* taking the shot?

To find out, change the odds slightly. Suppose that you have equal chances of dying from schmoo or dying from the schmoo shot. Suppose, say, a 1 in 100 chance of contracting and dying from the schmoo and an equal 1 in 100 chance of dying from the schmoo shot. Should you get the shot?

Scenario #2, equal odds:

There are several answers to this question. One answer is simply that it makes no difference. You could go either way. Death by schmoo or death by schmoo shot is still death. The outcome is the same in either case. The only appropriate decision on your part is to flip a coin.

A second answer is that it matters quite a bit what you choose and that you shouldn't get the shot. The reasoning may work something like this. If you get the shot and then you die, then you will have been partly responsible for bringing about your own demise. Such reasoning is fairly understandable. Supposing that you would not have died from the schmoo in a universe in which you had not gotten the shot, but that you do die in a universe in which you do get the shot, it seems that you can blame yourself for having taken an action that has brought on your death.

Conversely, suppose that you forgo the shot. Supposing that you would not die in a universe in which you do get the shot, but that you would die in a universe in which you forgo the shot, you can't as easily blame yourself for having brought about your death. As you lie dying in your bed, you can still lament, perhaps, that you should've gotten the shot, but you can't blame yourself for having brought about your own demise. At worst, you can blame yourself for having allowed your demise to be brought on. You let nature control your fate, much like you allow nature to control your fate in innumerable other circumstances.

Yet a third answer is that you absolutely *should* get the shot, because getting the shot will allow you to have some hand in the end of your life, which is preferable to a world in which you allow nature to take your life. One might argue that it is better to have brought on one's own demise than to have allowed one's demise to be brought on.

My point isn't that there is some line at which it becomes reasonable for you to get the shot over not getting the shot, or even that some people may have a higher threshold for risk than others. Rather, my point is that there is some important sense in which there is a nonutility value in me not having a hand in bringing about my own demise. Making these decisions isn't merely a matter of identifying the outcomes and understanding the probabilities. It is also a matter of understanding my responsibility and the nature of my responsibility.

EXPLAIN YOURSELF

If you're still not on board here, consider one final variation on this theme. Suppose that it is not you who will die, but a child you have never met. Let's call him JoJo. As a pediatrician—yes, you are a pediatrician now; presto, philosophy is magic!—you know that if JoJo dies as a result of something you've done, you will be forced to explain this occurrence to his parents.

In the event that JoJo dies due to the schmoo, which is a naturally occurring illness, you can rightly explain that JoJo has—terribly, awfully, horrifyingly—contracted a case of the dread schmoo. If, however, you vaccinate JoJo to protect him from schmoo, and he dies instead from the schmoo shot, then it seems to me that you have a fair bit of extra explaining to do. You will have to explain that you took the initiative to give JoJo the shot, acting in good faith that the odds would be better for him but not knowing whether he was a goo or foo carrier. You will have to submit that you did the best you could with the information you had, basing your decision on the original two-to-one odds.

Suppose, for the sake of this case, that you do decide to vaccinate JoJo, based on these original odds. Suppose further that nature is against you, that JoJo is more foo than goo, and so, immediately after your injection of his schmoo shot, he seizes up and dies. Had you left him alone, he would certainly have been fine.

Are you not at least slightly blameworthy for killing JoJo? His death is an immediate and direct result of your administration of the schmoo shot.

This, I think, is not the sort of vaccination program that we can straightforwardly endorse as a general policy, even though the odds are two to one in favor of the vaccine saving lives.

It's not that this sort of explanation cannot be offered. Certainly it can. I will happily accept two-to-one odds for myself and my son. It may also even be the case that the decision to give the schmoo shot can be justified for the entire population despite the risk. But if you offer this as *the* justification, what you will be doing, in effect, is offering a *justification by way of an explanation* and hoping that every party affected by this decision could accept your reasoning as valid. "I did it because it was more likely to help him than hurt him." It seems to me that it is less likely that JoJo's parents will accept your reasoning for the case in which JoJo dies of the schmoo shot than the case in which JoJo simply dies from schmoo.

Fine and well, you may be thinking; but if you hadn't vaccinated JoJo, then you would have been taking on an equivalent, if not greater, risk with his health. To argue this point, you may feel inclined again to cite the odds about the devastating effects of schmoo. But even if you do, there's a sense in which it's not entirely irrational for the parents to persist in their concern over your numbers. It is you who has done the vaccination, after all. You who have injected JoJo with the substance that activates his fooness. There is a direct causal line from your decision, the tip of your needle, to JoJo's deadly reaction.

The parents' threshold for risk will make an enormous difference in the ease with which you can explain yourself and on the extent to which they will release you from responsibility. If the span of odds is extremely narrow, say 1 in 100,000,000 people die of schmoo annually whereas 1 in 100,000,001 people die of the schmoo shot, your justificatory challenge is going to be mighty difficult. Many more parents may protest. If, however, the span of odds is quite a bit wider—if 1 in 10 kids is dying of schmoo, say, but 1 in 10 million is dying of the schmoo shot—the justificatory challenge will be much lighter. Depending on the parents' threshold for risk, there is likely to be a very dense bit of gray area.

I submit that this gray area is fleshed out by our concerns over matters like our culpability. Risk proneness or risk aversion is not, as some economists and psychologists like to pretend, a hardwired disposition. Rather, it is a philosophical difference, a different *view* about when our responsibilities kick in.

What we should be concerned about here is your complicity in harming the boy. Where information about potentially catastrophic outcomes may well move some people to switch directions, if one is really concerned about complicity, even the strongest appeal to catastrophe will be saddled with some question about the avoidance of catastrophe. It is better and more rational to be skeptical of the facts that move you to act than to be skeptical of the facts that fail to move you to act.

To get at this question, we need input not from our epidemiologists and engineers but from our ethicists.

• • •

All this schmoo talk is really just an analogue for interventions at the collective level. Spraying pesticides to keep down mosquitoes, thus preventing malaria deaths, is a widely endorsed but also widely condemned practice. Without question, some people will certainly be spared the terror of malaria, whereas others will face neurological complications from the spraying of the pesticides. To spray or not to spray, that is the question. It cannot be fully assessed by appealing simply to the risks and consequences. The same goes for many environmental issues.

It's not the facts (or the information) that will be doing the work in these cases, but the reasoning. The discussion about the science, about the state of the world, is a distraction from the bigger question of practical reasoning. Both environmental hard-liners and environmental obstructionists repeatedly fail to understand this point; instead they are insistent on bickering over the facts, the odds, the predictions, and the models.

There is no better illustration of this than the climate debate, in which almost the entire discussion is dominated by disputes over the numbers. There is surely a role for scientists in combating climate misinformation, just as there is a role for doctors and epidemiologists to offer justifications for schmoo vaccinations. Many people make a range of rational and irrational decisions based on no data at all, instead deferring to gut feeling, or instinct, or trust, or faith, or something like that. There may well be an important role for great debaters and bloggers to shift the discourse away from (or toward) a focus on climate catastrophe; maybe there's even a role for those who would shift the policy debate to more aspirational questions, like the burgeoning energy economy. But I have to believe that part of the problem lies in a failure to understand how policy making or decision making should and maybe does occur. This isn't a failure of science or of political ideology. What, after all, can scientists say that will alter the perspective of those who discount science because of its supposed elitism? It's a clash of views about what counts as adequate or good practical reasoning.

In my very humble opinion, more information, better science, and more positive framing won't do anything to move the discussion forward. That is much more likely to result in further entrenchment. What's necessary is a full-blown reconceptualization of what we think qualifies as making sense, as being justified—and it will take a lot of work from us, scientists and nonscientists alike.

Fact is, there are an infinite number of natural hazards in the universe, many of which can be prevented by this or that intervention. I can purchase a slightly bulkier or more expensive bike helmet, reducing my risk of traumatic brain injury slightly or dramatically, but perhaps choose not to. When I choose not to, I only may or may not be making this decision based on the tradeoffs of cost or style. My decision is also partly contingent on how much I could forgive myself for having made that decision (i.e., how responsible I would hold myself if something went wrong) versus how much I deem the intervention to be intrusive, necessary, or a risk worth taking.

But one cannot prevent or anticipate everything, and it would be insane to institute a rule in which reduction of harm is the sole objective. It is a fair bet that when you go camping in the woods, you consider the risks associated with bear attacks and murderous madmen. Perhaps you even go to some length to protect yourself from these risks, such as packing bear spray or a gun.

The same goes for the geoengineering discussion. There are a plethora of reasons why someone may object to geoengineering, even if the risks of geoengineering are known. As it happens, there is a similarly wide plethora of reasons to object to vaccine injection or pesticide spraying, even if the risks are known. It depends on what a citizen's threshold for risk is. And this, I think, can be established only by impartially assessing the reasons that we have for taking some risks and not taking others.

If we move forward with projects to geoengineer the climate away from the mistakes of our predecessors, then the engineers of such a mammoth project will have to accept responsibility for the outcome. They (we) as a collective will be to blame. They (we) as a collective will have to own up to it. They (we) as a collective will really have to apologize to our children if we trash the earth for good. I think that's unacceptable. We need to do something about climate change, yes; but we need to do something that only *reverses* what we've done, not that puts us on an uncharted climate path, forever lashing our collective responsibility to a policy that cannot possibly meet with the agreement, hypothetical or actual, of the billions that it will affect.

One last thing. Maybe, given the prevalence of conditioning technologies, you think that we've *already* altered the climate in such a way that many humans and nonhumans are, and will be, affected by climate change against their wills. Most of those affected couldn't or wouldn't (or at least didn't) assent to the changes that we're experiencing now. Maybe because of this, we should be less concerned about what future generations can assent to and instead just focus on digging ourselves out of this hole. This is a compelling idea. But the way in which humans have altered the climate has been willy-nilly. It hasn't been deliberate. Billions of people have acted independently, according to their own interests, to force the climate away from stability. Anthropogenic climate change is a colossal tragedy of the commons, a major failure of governance. We can't point the finger at any one individual, or even at any very large group of individuals, and say that they've done something impermissible or disrespectful.

In contrast, geoengineering is very deliberate. For us to move forward with a technology that will orchestrate and steer our climate away from this tragedy, to a—fingers crossed—better outcome, is not simply for us to *act* on our planet, but to *react* to the negative effects of an uncoordinated and chaotic multitude. It is to accept the tragic transformation of the climate and to patch it over with a collective bandage. It is to inject climate change with our collective culpability. Whatever happens *after* that point—once we have dramatically altered the flora, fauna, and chemical composition of our oceans—we *will* collectively be to blame.

But if we're not to envision or picture an ideal world and then strive toward that, then what? How are we to be environmentally responsible, how are we to build green businesses, if we don't have a vision? To my mind, the answer lies in what we have been discussing in the past several chapters. We must take a deep breath and ask ourselves about our choices. We must ask about each and every choice, about the conditions under which our decisions could be said to be justified.

Whenever possible, we should keep our options open. As we devise our business plans, we need to ask what our reasons are and whether we can justify our business to a community of not so like-minded others, to those who stand neither to gain nor to suffer under our eventual action. If we aim to create a product, we must ask whether others would agree that the product is important. If we advocate a project or a policy, we must ask about the reasons that support it: Is it doing important work, and can it be done in the right way? It's not enough for us to allow the market to decide. It's not enough for us to allow individuals to make the decision for themselves. Consumers have all kinds of screwball desires, and consumers can be easily manipulated. We must ask collectively whether our products, programs, or projects serve a real purpose that could and should be aspired to. We need to have a serious debate about our reasons.

7

The Voter's Conundrum

In which the author (1) raises concerns about why people vote, comparing these to sundry other environmental issues, (2) explains that the broad spectrum of motivating reasons reside in our will, (3) examines the extent to which explanation and justification become intercalated, (4) breaks a few eggs, (5) gets drunk, and (6) offers a view of the right as the justified.

SUCKER'S BALLOT: WHY WE VOTE

In 1968 Paul Ehrlich warned that if the alleged "population bomb" were to explode, hundreds of millions of people would starve to death in the 1970s and 1980s. Less than 30 years after the publication of his dire prediction, the world population essentially doubled, resulting in nonnegligible bouts of starvation, but nothing nearly as devastating as the erstwhile town crier had prognosticated—unless, of course, you happened to be one of those unlucky starving people, in which case you would've experienced your own private and particularized devastation, slowly and painfully languishing while cherry-cheeked humans in other parts of the wealthy world, the great Ehrlich included, were heartlessly and happily gorging themselves on fried chicken, chocolate ice cream, and hollandaise sauce, though none of those ingredients, god forbid, all at once.

As the people-ticker tocked and the farming-tocker ticked, the US population grew from 200 million to 300 million, very few of whom would starve to death and very many of whom would have done quite well for themselves. The period from 1960 to 2000 were the glory years for American prosperity. The US GDP nearly tripled, the average house size expanded from approximately 1,500 square feet to 2,500 square feet, and the waistlines of Americans ballooned to match.[1] All of which, I am embarrassed to confess, has very little to do with the ruminations of this chapter, except insofar as *The Population Bomb* trades in unfathomably large numbers.

By 2008, 40 years after the Paul Revere of Population rang the people alarm, roughly 225 million of the 300 million Americans were 18 years or older. In that same year approximately 58 percent of the US voting-age population, nearly 131 million people, headed to the polls and pulled the lever either for US presidential candidate Barack Obama, John McCain, or one of several other fringe and forgettable candidates.[2] The turnout was heralded as a record success, due in part to the appeal of African American candidate Obama. Pundits around the country saw the turnout as the dawn of a new day in American politics; the day when minority populations no longer felt disenfranchised and misunderstood and the youth finally grasped, once and for all, that their piddling bit of ballot punching actually mattered to the outcome of the election. November 3, 2008, was therefore one monumental day in history, on which extraordinarily teensy actions—the checking of a box, the pulling of a lever—were said to have extraordinarily big consequences.

All of this might well have been anticipated, for it had been only eight years earlier, on a similar such monumental day, in the very hairy 2000 presidential race between candidates Al Gore and George W. Bush, that individual votes resulted in the weeks-long Florida recount, winnowing state and national races down to double-digit leads. Most remember this event as one of great drama and intensity, as Republican activists banged down the doors of the Miami-Dade Canvassing Board office while scores of dedicated actuaries squinted and peered at a ridiculous taxonomy of chads, culminating in the allegedly bipartisan Supreme Court decision that would eventually determine the election and seal the fate of the United States for the next four (if not eight) years.

It is common in our political discourse to hear justifications for voting that operate along these lines. We are told that we ought to vote because every vote counts and every vote matters. The votes pile on top of one another like so many jelly beans in a giant jar. If a race seems like a slam-dunk for our preferred candidate, we are sometimes encouraged to go to the polls anyway, since we can't always count on our neighbors to pull their weight; and besides, more votes will mean a stronger mandate, so we need to show our support for this reason, too.

Under most circumstances it is simply not true that our individual vote matters. If the point of an election is to win, then, mathematically speaking, it matters little whether a candidate wins by one vote or 5 million votes. Slogging to the polls and standing in line is, on this view, a waste of your time. It doesn't make sense for you to spend a good portion of your time researching candidates, learning about their positions, and eventually voting. It's irrational, at least if what it means to be rational is established by how efficiently you allocate your resources and choose to spend your time. You're expending an enormous effort for what might be no reason. Nevertheless, more than 130 million people vote in US presidential elections.

Could 58 percent of the US population be this irrational? Is it possible that the remaining 42 percent of nonvoters are actually the smart ones, keen to the false sense of contribution that voting offers?

I think that those who vote are actually quite reasonable in voting, though I also agree that their vote might not count for much. Put a little differently, I think that the reason we should vote has little to do with whether our vote will amount to a larger or smaller victory. I'll explain why in a moment.

Many environmental problems ultimately are much like voting: no single action has any noticeable effect. Using one less sheet of toilet paper will probably not make a difference in the overall destruction of forests; one more plastic water bottle does not substantially increase the impact on landfills; one shorter shower doesn't make for an abundance of water that we can use to quench the thirst of dehydrated crops; cutting back on in-home toxins will probably have no measurable effect on the concentrations of dioxin in the water supply. For almost any environmental issue, the nominal action of any given individual won't make much difference in the aggregate outcome.

It is not unreasonable to claim, as some prominent public figures and philosophers have claimed, that taking a drive on a Sunday afternoon has no effect on the climate.[3] In fact, a full lifetime of individual carbon emissions doesn't really push the climate in one direction or another. And yet it is demonstrably true that every single polluting action, stacked on top of all the other related polluting actions, has an aggregate effect no more poignantly illustrated than in the case of atmospheric carbon concentrations, which in 1880 were approximately 280 ppm but now stand just above 400 ppm. As we innocently live our lives, atmospheric carbon concentrations rise and rise.

Such circumstances are typically treated as one of several variations on the famous *tragedy of the commons*, a view about environmental harm that was popularized by biologist Garrett Hardin in exactly the same year that Paul Ehrlich was blowing his stack about rampant population growth. The familiar idea is that individual actors, when acting according to their own self-interest, in a circumstance in which there is little opportunity for oversight or enforcement, will paradoxically act *against* their own self-interest by depleting the very stock on which they depend. A shepherd with a flock of sheep and access to a common grazing area—that is, open to a bunch of other shepherds and sheep—may inadvertently overgraze the area, thus jeopardizing future grazing of his or her sheep. The shepherd, in doing what was thought best for the sheep, will paradoxically have undermined the ultimate survival of the flock.

The diagnosis of this problem as the tragedy of the commons gives rise to related prescriptions. If, for instance, the depletion of blue crabs and oysters in the Chesapeake Bay is attributable in part to the unsustainable practices of local crabbers, then the correct thing to do is one of the following: get the crabbers to see that their individual actions, when taken in concert with the actions of the others, have an aggregate impact that will be detrimental to their self-interest; get them to see that endorsing a policy of cooperation between all crabbers will serve each one's best interest; or, more paternalistically, regulate and monitor their behavior so that all the crabbers comply with a law designed to keep stocks high by preventing cheating.

This prescription for action isn't quite as straightforward when the beneficiaries don't bear the burden of the consequences of their behavior. If the downstream negative nitrogen balance of the Mississippi Delta, say, is due in part to the upstream farmer's fertilizing of her crops, the farmer may not have a clear reason to refrain from fertilizing her crops. Fortunately, there are policy kludges for this, too: give the downstream parties the capacity to sue for damages by assigning enforceable property rights; this will give upstream offenders some incentive to avoid causing downstream harm.

Unfortunately, regulation and monitoring become more difficult as the number of stakeholders grows. Global problems, like climate change and the depletion of ocean fisheries, tend instead to appeal to the moral conscience of actors. For those who subscribe to moral theories that propose that the right thing to do can be circumscribed by appealing to the best state of affairs (i.e., the outcomes), it is tremendously disheartening to think that one's actions won't have any effect at all. Mammoth-scale tragedies of the commons, like climate change, seem to suffer from this problem. If a leisurely drive on a Sunday afternoon will have no significant or noticeable effect on the climate, what's to militate against it? If even a full year's worth of emissions will barely make the needle move on atmospheric carbon concentrations, why should anyone cut back? Why should you be bound, morally speaking, to reduce your emissions if no action you take is going to make a lick of difference? Indeed, if the right act is established by whatever brings about the most good, and one's action won't have much of an effect on bringing about any good, then this is a mighty disempowering principle indeed. There is no right action.

To my mind, however, there is an important sense in which this logic elides the ultimate reason that individuals should be cutting back on their overfishing, their use of toilet paper, or their overapplication of fertilizer. The logic may indeed explain many cases of commons tragedies, but it is woefully insufficient to justify the prescriptions that allegedly arise from these tragedies.

In chapter 5 I introduced you to motivations, and in chapter 6 I suggested that our motivations alone are not adequately characterized in terms of means and ends. Now I want to return to that discussion of reasons, but this time focus carefully on the distinction between our *motivating* reasons (the reasons that move us to act) and our *justificatory* reasons (the reasons we use to justify our actions). I want to carve justificatory reasons more precisely. What I'll try to suggest is that much of the time we are motivated by our *brute* or *animal reasons*, which are tightly intertwined with our wants and desires, whereas we ought to be motivated by our *normative reasons*, the reasons that give impartial moral and ethical weight to our justifications. I want to investigate in somewhat greater depth what those normative reasons might look like.

The first third of this book was structured primarily around explanation. I explained extinction and the climate debate, introduced the "precious vase" view, and ultimately offered a few examples of just how awful and terrible nature can be. In the course of these explanations, I explained explanation. I basically told you a story that has "truth value," with which you can agree or disagree, and that I believe is narratively compelling but can be challenged on truth grounds by historians, anthropologists, sociologists, and political scientists.

The second third of the book was my attempt to draw you away from explanatory narratives to contrast actions with events. In particular, I hoped to direct your attention toward the role of justificatory reasons in distinguishing actions from events. From here on I'll focus more directly on justification.

WHERE THERE'S A WILL, THERE'S A WHY

"Would you like another beer?" the waitress asks you politely. You've been saddled with this dilemma many times before. Yes, of course you'd like another beer. Beer is nice; it makes you feel good, it tastes great, you're out with your friends having a good time, and so, without thinking, you may be inclined to answer yes. But then it hits you. You're a little tipsy already, you have to drive home, you need to be alert at work tomorrow morning, you're trying to lose weight, and you have a niggling feeling that you should probably be home in time to watch your favorite program with your partner for some much-needed snuggling.

In this case, the bar owners hope that you act according to your brute desires and just do what you want. To give the waitress your answer, however, you must travel back through your other commitments, to the multiple conflicting reasons you have for not grabbing another beer. Once again, your will finds itself in tension with your wants.

It is perfectly understandable that some may straightforwardly conceive of this decision-making process in terms of the intensity of wants, in terms of a stronger desire that overtakes all of your other desires. There is an incredibly rich tradition in many branches of philosophy, and certainly in modern economics, of characterizing your dilemma narrowly in terms of desires. What these philosophers and economists would have you believe is that given more information and more time to think about your decision to have or not have the beer, your answer will be that you do not, in fact, "want" the beer. I take it that this is the accepted view: that all *oughts* can be understood in terms of our desires.

Someone might believe that this is a robust model because, all things considered, there are other things you want or desire *more than* the beer. The reason that they say this is that you have other conflicting desires. You want to feel normal for work tomorrow, you don't want to get arrested for drunk driving, you don't want to become tipsier, you want to be thinner, and you want to fulfill your promise to watch television with your partner. Given all your other wants, it can be easy to think that the appropriate answer to this question is one of lesser and greater wants, each of which outweighs another to the point that you arrive at what you really want. I think this is a funny way of thinking about wants, but I definitely see what's attractive about it.

Consider the dilemma in terms of smiley-face satisfaction units. Maybe you'll get five smiles from a beer but ten smiles from snuggling with your partner. (Philosophers sometimes speak of *utils* or *hedons*, but units such as these can be deceptively slippery because they're not necessarily tied to the satisfaction of desires; or if they are, it is through sometimes convoluted connections between what we know we want and what we would want with more information about our wants. Here I'm speaking only of known desires.)

Beer = ☺ ☺ ☺ ☺ ☺
Snuggling = ☺ ☺ ☺ ☺ ☺ ☺ ☺ ☺ ☺ ☺

In this smiley-face count, snuggling is far preferable to another beer. Yay for snuggling!

As I've said, however, I think this is a funny way of thinking about wants. It's perfectly reasonable to want both things, whether equally or unequally, but only to be able to have one.

It also seems perfectly reasonable that we might find ourselves with an arrangement in which we stand to gain more satisfaction from the beer—in other words, we want the beer more—but instead opt for snuggling.

Beer = ☺ ☺ ☺ ☺ ☺ ☺ ☺ ☺ ☺ ☺
Snuggling = ☺ ☺ ☺ ☺ ☺

Yay for beer! I'll have the snuggling, please.

You might, for instance, decide to head home because snuggling has been a long time coming and you feel a strong commitment to your partner. Perhaps your commitment to snuggling can be expressed in a greater number of smiley faces, but there are reasons to suspect that these smiley faces aren't directly translatable into other units of utility. The reason that you will go home to snuggle is not because it makes you happier, but because *that is what is sometimes required* in a committed relationship. That is what you *do* in a relationship. And despite all this, you may do it without begrudging your partner. Maybe you still get quite a bit of satisfaction out of snuggling (e.g., five smiley faces), but not as much as you would from beer.

Hardened economic types sometimes have a tough time seeing this, but the point isn't that one option is better or worse than the other; the point is that the relationship gives you a reason that can, should, and often does guide your behavior. It might be a reason that gets in the way of the maximization of your satisfaction, but it is nevertheless a reason that is uniquely yours. You might think of it like this, where the wedding bell indicates the intensity of commitment to a partnership:

Beer = ☺ ☺ ☺ ☺ ☺ ☺ ☺ ☺ ☺
Snuggling = ☺ ☺ ☺ ☺ ☺ 🔔 ⟶

We make commitments like this all the time—at our jobs (butchers, bakers, candlestick makers), in our family lives (fathers, mothers, sons, daughters), in our social lives (friends, Romans, countrymen). These commitments count as reasons too. Sometimes we are born into relationships with deeply embedded and obscure commitments—we are members of a family, or we associate with a particular section of the country—and we might have to isolate and identify these embedded commitments. Sometimes we accept, adopt, and endorse commitments one by one, such as when we take out a mortgage or make a promise. Sometimes we must embrace extremely complicated relationships, replete with good and bad commitments, because we are particularly drawn to one or the other principle, such as when we propose marriage or find a political party to our liking. Whatever the nature of these commitments, it is critical for us to make sense of them since they too count as reasons.

Now then, you may feel strongly that these commitments are still translatable into a hierarchy of wants. Perhaps you think as many of my students think: these commitments can be characterized in terms of short-term gratification versus long-term gratification. Yet this can be shown to be clearly false. Sometimes our relationships give rise to reasons that won't be good for us in any respect. Maybe we choose to forgo a beer so that we can visit our sick and comatose grandmother in the hospital, even though we find it emotionally difficult to see her in the state she's in, and even though she will probably never know that we've come to visit her.

Beer = ☺ ☺ ☺ ☺ ☺
Grandma = ☹☹☹☹☹☹☹☹☹☹

Sniff. Sad for grandma. There is no long- or short-term gratification in visiting Grandma. You visit her because you should.

By endorsing or rejecting commitments, or by forming relationships that play host to such commitments, we fashion for ourselves what Harvard philosopher Christine Korsgaard calls *practical identities*. We carry our practical commitments around with us and filter the world of likes and dislikes, wants or desires, through these identities. I visit Grandma because I fancy myself to be a good grandson; I forgo a beer because I understand myself to be a good husband.

If we unreflectively accept an identity because that is how we were raised—maybe we were born on a hippie commune or bred a Southern boy—we effectively allow the identity to govern our behavior. If, however, we adopt and endorse an identity—such as a militant vegan or a gun rights advocate—through reflective engagement with all that it entails, then we can rest assured that our advocacy comes from somewhere personal: our will. We can say, in effect, that we are drivers, self-legislators, equipped with the steering capacity to act for normative, and not brute, reasons.

None of this is to say that the identities we assume are therefore justified. It is only to observe that these identities and relationships serve as host to our broad spectrum of underlying reasons. On one hand, the norms that govern these identities can go a long way in explaining why we do one thing and not another. On the other hand, once we have isolated these embedded commitments, these norms, we can then position ourselves to better understand the rightness or wrongness of them. In this case, the reasons serve double duty, both to motivate and to justify: they motivate us to act, and they serve to justify the act. This is vital, since the question that most ethicists want an answer to is *why one norm is better than another*, or why one norm justifies our behavior better than another norm.

Fact is, our identities, particularly those that we endorse and adopt, are packed to the gills with normative reasons that frequently override the pull of our wants and desires. Ask an environmentalist why she so identifies with the movement, and she is likely to rattle off any number of reasons similar to those that we explored in chapter 3. These are reasons that she has *taken up* for herself and made a part of her identity. She was likely not born into environmentalism. Instead, she believes, rightly or wrongly, that these reasons *justify* her stance. Moreover, abiding by our Super-duper, double-secret, underlying principle of humanity, these reasons that she has reflectively endorsed and adopted are *in principle* justifiable. These are the reasons that guide her; and it is these reasons, not the simplistic motivational reasons that explain her behavior, that we must scrutinize.

Unfortunately, our beliefs and desires are the deepest and most familiar trove of reasons for us to mine. We feel our desires deeply, and we know our beliefs well. Social scientists, policy makers, and bartenders are quick to excavate these beliefs and desires for their personal gain, happy to explain why we're doing what whatever it is that we're doing. In most cases, this is fine. Many of these reasons serve double duty: both explaining and justifying our behavior. As an adult, my justification for drinking a beer simply *is* that I desire a beer. I don't need to do much else to justify drinking a beer. If I were 10 years old, my desire for a beer would have to be measured against a wider set of considerations and complications that make the fulfillment of such a desire problematic.

At any given decision juncture, we face the possibility of allowing our desires to drag us along or of acting for reasons that we take up and endorse. The justificatory reasons that guide our actions are norms, principles, values, commitments, obligations, and responsibilities that all, in one way or another, answer the question *why*. Why did you do X?

EXPLIFICATION

US political scientist Elinor Ostrom surprised and shocked the intellectual world when in 2009 she was awarded the Nobel Prize in economics. Not only was she the first woman to receive the award, she was the first noneconomist.[4] Her work, which she conducted with Oliver Williamson, explored the governance of commons issues. It suggested that humans interact with ecosystems differently from what is anticipated by traditional neoclassical economic models, such as Garrett Hardin's, but that economic models can apply to other institutional arrangements as well.

Ostrom found her foothold by conducting economic research on institutions other than markets—firms, associations, households, agencies—which was indeed an innovative approach to understanding the inner workings of these non-market institutions. Essentially, she challenged the economic orthodoxy by suggesting that networks and communities play a vital supporting role in the organizational efficiency of markets. Her findings are fascinating. They break new ground and shed light on issues that many economists had otherwise taken as a foregone conclusion: namely, that sometimes actors in these institutions behave irrationally.

I think it's probably fair to suggest, however, that Ostrom's findings are as robust as they are because people are frequently motivated by nonutilitarian considerations, as we saw with the schmoo vaccination and in the beer example. The obfuscating factor, I suspect, stems from confusion over the nature of reasons, a confusion that loops all the way back to chapter 1, over the nature of answers to *why* questions; a confusion that implies the deficiency of playing sucker to the idea that human beings are in some respect *rational*.

Consider it this way. The infamous bank robber Willie Sutton was once asked by a reporter, Mitch Ohnstad, why he robbed banks. "That's where the money is," he quipped. Philosophers and writers love this response. They love it so much that I considered avoiding it entirely. It's concise and straightforward, and it's also painfully clear how handily it derails the questioner. The gag works because it's obviously a good answer to the wrong question, spoken by someone who deliberately misunderstands the question put to him. Why else would you rob a bank?

What is clear here is that Sutton plays on an equivocation in the question *why*. What's a little less obvious is that this equivocation is embedded in every *why* question, and we're often just as silly about answering *why* questions.

If Sutton had answered that he needed the cash to support his dying child, he would have given us a somewhat better answer—a start at a justification, perhaps—that would help us make sense of his action. We might even have been prepared to forgive him. What he would be doing is offering us a reason that we might accept or reject as valid or strong. If he had done this, he would have injected his response into a completely different conversation—about the permissibility or impermissibility of robbing banks in such-and-such a situation. He would have shifted the focus from an explanatory stance to a justificatory stance, appealing to others to accept his reasons for robbing banks. He would have endorsed this reason as his core reason—motivational because it moved him, justificatory because he thought it justified—for all to scrutinize. He would have offered the rest of the world an opportunity to weigh his reason against the other values, principles, constraints, and considerations to which we are all subject.

It's not uncommon to find people who, like Willie Sutton, are fixated on understanding problems in terms of what offers the best explanation. They rob banks because "that's where the money is." Many people who answer questions this way are not smart alecks, and they often don't even realize that the answers they offer don't actually get at the question that is being asked of them.[5]

Ostrom's findings, for instance, sought an answer to a *why* question. Why do people in nonmarket arrangements sometimes find successful resolutions of commons problems? She found her answer in the customs and conventions that facilitate cooperation. This is a hugely helpful explanatory perspective through which we can make sense of why people act the way they do. At the same time, it is hard not to feel that there is something lacking in the answer. It does not anticipate or parse why those norms and principles have enough compelling motivational force to be taken up by her subjects.

Ostrom's research answers why people do what they do by explaining *that* they do what they do. It does not offer a clear answer to why those norms and conventions prevail. Surely there is a speculative answer to which she defers: maybe it is better for those involved in small manageable commons-style situations to arrange shadow conventional systems of rewards and sanctions to keep actors in line. But there are many other such speculative explanations that may serve equally well: maybe the norms propagate through memetics, a sort of evolutionary genetics of ideas; or maybe the norms have been established by the forceful fist of authority, by the opiate of magic and superstition, or via the decades-long cultural replication process of traditional dance.

These alternatives are all just explanations of the propagation and perseverance of ideas. The correct answer to the *why* question can't be reached through these explanatory channels. Instead, it can be reached only by deference to the reasons that ultimately guide the actors and to the inferential strength that connects one reason to the next.

Many times, answers to *why* questions are hybrid explanation-justifications: *explifications*. They are explanations because they offer up a conclusion about how best to bring about a certain state of affairs. "If you want to make people happy, you should do X." But they masquerade as justifications because they suggest that an act is right if it brings about the greatest good, whatever that good may be. "Freedom isn't free." These aren't true justifications.

Many people who study these things, Elinor Ostrom included, are inclined to answer *why* questions with explanations. They explain why people behave as they do; for instance, people in small groups sometimes act according to institutionalized norms. This can be a hugely helpful way of understanding what's going on, so I don't want to disparage it. Ostrom rightly deserves the Nobel Prize for her work.

Other people, however, like my colleagues in philosophy, lean toward a justificatory response. When I ask them why, they launch into an elaborate investigation of what serves as the best normative reason: it's not just that people sometimes establish and activate these norms, or even that they may do so out of self-interest, altruism, impartiality, fairness, consistency, or some other unspecified goal; it's that it makes sense, or it doesn't make sense, for them to act this way.

Justifications of this sort, normative justifications, come in at least two varieties: justifications that appeal to the good and justifications that appeal to the right. Justifications that appeal to the good generally assign or uncover a value in the world. In ethics, the good of note is very often happiness, but it can just as easily be health, money, freedom, nature, or even Elvis memorabilia. An act is justified if it promotes X, where X is any one of these goods.

Thus, some people say that an act is "right" because it brings about the greatest amount of good, such as happiness (i.e., it makes the most people happy). It's a pretty common way of thinking. In our practical lives, these conceptions of the good vary from person to person, from locale to locale, and yet they are always lurking in the background of our political discourse. Frequently they go unstated; or if they are stated, they are clumped vaguely into the black box of "values," not to be mentioned in polite company. Some philosophers spend a good deal of their energy arguing for one of these conceptions. Unfortunately, thinking like this leads many to assume that they will have to make some necessary trade-offs—jobs versus nature, for instance; or that we can characterize all value propositions in terms of one monolithic good, such as cost versus benefit.

Justifications that appeal to the right, in contrast, tend to emphasize obligations and commitments as having priority over any particular conception of the good. In these cases, some will say that an act is "good" if it honors or respects the right. So sometimes, as we've seen, we may have reason to override our personal happiness, or maybe even the happiness of others, to fulfill some prior obligation. So too with health, as we saw with the schmoo shot. The question for those philosophers who appeal to the right is not what good is the best good, but *how we derive these obligations and commitments*. How do we know that these are the correct standards?

There are certainly very different answers to this question: some say that moral commandments have been handed to us by a divine being; others deny altogether that there are any absolute obligations and instead insist that they all boil down to mere convention, to codes of etiquette. Indeed, if you're on board with what I've said above, you may think just this: that our commitments are little more than social conventions directed primarily at helping us achieve our ends. I think that's false and that moral reasons can be wedded tightly to our capacities and commitments as reasonable and rational agents.

Moreover, you may be concerned that the right-reasons approach I'm outlining here also suffers from the problem of trade-offs, just as the do-gooder approach does. Even though one may not be forced to decide between conflicting values, one has to decide between conflicting reasons instead. But I think that's false. I think we have a way of sorting it all out: a decision-making procedure, if you will, that establishes which acts are reasonable or acceptable or justified and which acts are unreasonable or unacceptable, or unjustified. Allow me, please, two non-environmental illustrations.

WORD-SALAD SANDWICH

An omelet is a flat layer of scrambled eggs cooked to look like a crepe. In America, it's shaped that way in part so that you can put lots of great stuff in it, from cheese to tomatoes to beef jerky to chocolate chips, before you fold it over on itself. In France, an omelet is a more delicate endeavor, folded over on itself twice and often served plain or with fewer ingredients. When done right, both are amazing, provided that you're down with the filling.

Most of us have intimate experience with this breakfast delicacy, and many of us have probably mangled a few dozen (or in my case, a few hundred) omelets over the years while trying to impress friends and family on Saturday mornings. I botch approximately two omelets for every three I make.

What's interesting about an omelet is that it falls into a category of breakfast dishes that are also displayed on the menus of most diners and restaurants alongside a vast range of other egg dishes: fried eggs, poached eggs, scrambled eggs, soft-boiled eggs, and even shirred or coddled eggs. Typically, we see these options on a menu and order them as they're written: "I'll have two poached eggs, please." What we're doing here is asking for a dish, two poached-eggs, the same way we might ask for an omelet: "I'll have an omelet, please."

This is curious, because what we're actually referring to in many of these cases is not the egg dish itself but rather the way in which it came into being. A poached egg is an egg that has been poached, which is a technique for cooking. A fried egg, of course, is an egg that has been fried, another cooking technique. With the listing of the dish on the menu, the technique of cooking the eggs, or the process by which the eggs acquired their name, fades into the description of the dish. This is a straightforward way of understanding the classic options.

Right beside these possibilities are a variety of egg dishes whose names do not include the cooking technique. Along with omelets one can usually also order eggs benedict, eggs florentine, huevos rancheros, eggs connaught (whatever that is), and an assortment of other mouth-watering breakfasts. It makes sense that we order these eggs according to their dish description and not according to the technique by which they're cooked.

We order most of our foods by describing what they are and not how they're made: "I'll have the chicken parmesan, caesar salad, and mushroom risotto, please." More colloquially, we might shout to the second-order cook, "Adam and Eve on a raft, wreck 'em," and expect that two scrambled eggs on toast will eventually find their way to our table. It's a culinary curiosity when we distinguish a plate of food not by the name of the dish but by the technique: "I'll have the salmon poached in white wine, please." Just as with the omelet, we can offer an adjectival description of the egg dishes that characterizes their genesis.

This discussion may sound a bit frivolous, but I assure you that it is not. When we talk about an act being justified (or reasonable or acceptable), we are subject to exactly the same confusion as when we order an omelet versus a dish of poached eggs. We refer to the act description ("That act was justified") and not to the technique, method, or process of justification ("That act, or the reasons guiding that act, went through the requisite process of justification").

The label *justified* is frequently pinned on acts like a badge even when they haven't gone through the requisite process of justification. We are, as well, subject to the same confusion when we talk about convicted criminals. We use the term *convicted* as an adjective describing the noun *criminal*. The criminal wears his conviction like a badge; his description as a *convicted* criminal distinguishes him from an *accused* criminal, much as the description *poached* egg is distinct from *fried* egg. We might as well be saying that they are violent criminals or nasty criminals, or even blue or chartreuse criminals. Just as with justification, just as with eggs, this isn't always what we mean. What we mean is that the person in question has been convicted of a crime, and this conviction has occurred through an elaborate process involving the accumulation of evidence and a trial with lawyers, witnesses, a judge, and a jury. This is *due process*—a process due the person so that the principles of fairness, justice, and liberty are respected.

We can't call a criminal a *convicted* criminal unless he's actually gone through this process because his conviction is dependent on the process itself. Similarly, we shouldn't call an act *justified* unless it's gone through the requisite justificatory process. It is not enough to call an act justified simply because it meets with our personal (and likely misguided) idea about what a good state of the world is. Take the word *guilty*, for instance. Randall Lee Smith was guilty, in the technical sense of the word, the moment he pulled the trigger; but in another sense of the word—in the eyes of the state and in the eyes of the law—he was innocent until proven guilty. He wasn't guilty until the judge pronounced him guilty.

If I fail at making an omelet and instead serve a mangled platter of eggs, there is a sense in which I am serving, in fact, the same dish I would have served had I scrambled the eggs; but when I serve such eggs, I find myself insisting that they are not scrambled eggs, that they are, instead, a mangled omelet. The reason for this is natural: I had intended to make an omelet, but that process was thwarted by my incompetence, so I am serving, instead, my mangled omelet.

So too with the moral standards by which we live. We must go through the arduous and ongoing process of excavating the norms and standards by which we live and ultimately justify them by subjecting them to the scrutiny (or at least ensuring that they *could* be subjected to the scrutiny) of outside, disinterested parties. If perhaps we stumble on the right and sustainable lifestyle without undergoing this process, we will only have *stumbled* on the right and sustainable lifestyle. Though most of us likely take many actions that we consider right and justified even though they have not been put through this arduous justificatory process, I suspect that this is because we have become very efficient at taking short-cuts through the justificatory jungle. We do not need to ask our neighbors if it is wrong to shoot their dog, because murdering a family pet is pretty much a settled matter. But if we were to shoot the dog—say, because it is attacking our child—there is an important sense in which, if called upon to justify our action to them, we would feel the need, we would feel the pull, to offer up our reasons for doing what we'd done. "I'm very, very sorry for what I've done, but given that your pet was gnawing on my son, I had to take this extreme and unfortunate action."

OF WHISKEY AND DARTBOARDS

In every one of my classes I am required to give grades. As is standard in the United States, I grade on an A through F scale. Almost always, however, I engage in some degree of grade inflation. I also engage in strategic grade giving. As a rule, I score my students more harshly near the beginning of the semester than I do near the end of the semester. I do this primarily for pedagogical reasons, to strike fear in their tender little hearts, to inspire them to hit the books as they've never hit them before.

Like many professors, I have strong feelings about grading. Much of the time I feel that it is the worst part of my job, though as far as worst parts of jobs go, it's really not so bad. To take the edge off of really long, late-night grading sessions, I sometimes joke with my students that I will force myself to power through the grading by cracking open the whiskey cabinet. My students get a kick out of suspecting that my altered state may help them fare better on their papers, as if I am drinking to the point of annihilation, like a latter-day Don Draper. Truth is, I'm more of a morning grader, so most of my grading is done beside a stiff pot of coffee. Several cups of that potent intoxicant, and I'm ready to chew them to pieces.

Apart from the insane monotony of reading poorly formed essays and half-baked thoughts, there's the question of what metric I should use and what, exactly, I am to do as a grader. Inevitably I face a conundrum.

I am torn between penalizing my students, on one hand, and giving them important and helpful feedback, on the other hand. With the strong students I run the risk that my praise will give them an overinflated sense of their competence. Perhaps they will stop working so hard. With the weaker students, I run the risk that my criticism will be counterproductive, either turning them off entirely from philosophy or forever marring their records. Maybe they won't get into grad school, for instance, because the score they've gotten is too low. Maybe they won't even aspire to go to grad school because I have so persuaded them that they are quivering numbskulls. Frankly, any given philosophy class isn't worth that kind of long-term marring. Usually I find a middle ground. Such a conundrum should be familiar to anyone who has had to offer an assessment or a judgment of anything.

Fact is that most undergraduates, even the very smart ones, aren't terribly good philosophers. They're better at philosophy than high school students, to be sure, but that's a maturity thing. I could spend hours in class helping them see where they're mistaken and how much they have to learn, lasering in on the minutiae of philosophy, hoping to raise their work to the level of my colleagues, but doing so wouldn't be constructive. Rather, to evaluate them I have to make informed judgments about what they're capable of and where they are in their development.

In principle, a grade is an evaluation, or a measure, of a student's performance. As such, it is often the case that depending on the grade, a student's understanding of what the grade is measuring will shift. What I usually tell my students, several times a semester and almost every time I hand a grade back to them, is that the grades are a method for me to communicate with them what I consider the quality of their work to be, with all relevant caveats about their capabilities, background, and experiences taken into account. I may grade an essay very high if I think that the student has tackled a particularly tricky issue, even though she has not done so with the kind of rigor that I would expect of my colleagues. I may grade another essay very low if the student has expressed an interest in pursuing this issue in a thesis or dissertation or in a future publishable article. All these factors go into my communication of my assessment, of my judgment. The grade is just a very clumsy way of communicating this judgment.

I don't claim to be the ultimate authority on their work. I don't expect that they'll agree with every one of my assessments. I just want to convey to them my professional judgment. As it happens, this becomes all the more important as we get nearer the end of the semester and I have to make a determination about their final grade. In this case, the university confers upon me the authority to pronounce their work of a certain quality. By virtue of my station, by virtue of my professorship, my assignment of a grade effectively cements my evaluation of their work for all eternity as of a certain quality. And this is critical. Like a judge, when I give someone a C, I effectively transform that student from a mere student into a C student. The student becomes, through the act of my anointing him so, a C student.

This is not to say that before I offer my grades my students are not C students and that after I offer my grades the students become C students. That's not how it works. The students are, presumably, C students even before I give them their grades.

The important point through all of this is that to arrive at my conclusion about the final grade that each student deserves, I have to go through a considerable process. I read through everything. I look at it once, twice, sometimes three or four times. I stack all of the student's documents in a pile beside my grade book and make an assessment by using my judgment. It is, in many cases, a rough assessment, but I try to be as fair as possible. As I do this, I write down my reasons. For instance:

> Roger's paper lacked a clear thesis. Sarah did a really fine job on her presentation. Josephine didn't participate in class as often as she should have. Several times it was clear that Demetrius didn't understand the material. Cody took us on a tangent.

This is a time-consuming project, but it's not unlike the process through which workers are scrutinized in their annual performance evaluations. Indeed, I try to explain as much when I inform them of their grades. "This is a dry run for the real world," I say. "In the real world, if your writing is terrible, you won't last long."

Given the onerous burden of grading, I have considered assigning final grades through other, totally irrelevant, means. I might grade them based on their appearance; for instance, A for the well-dressed, F for the piggy kids. Or I might assign grades by shooting darts at my grade book. If I were to do this, there is a nonnegligible chance that I would one day hit the grade book in exactly the right way, corresponding to the same pattern as the grades that the students actually deserve.

Suppose just such a thing. Suppose there is one student, Tina, whose work I feel too tired to scrutinize. It has been a long day, and after hours of reading incoherent drivel, I determine that there must be a better way. In lieu of reading her papers, I simply decide that I will drink whiskey and throw darts. I set up an elaborate method for doing this. My dart strikes the letter B. Tina will get a B in the class.

Suppose further that if I had read through Tina's materials, Tina would have received a B in the course anyway. Since I did not read through Tina's materials, but instead chose to throw darts at a dartboard, I assign Tina a B based on my exceptional dart throwing. Now then, was Tina's grade justified even though it resulted from throwing a dart?

On one way of construing things, Tina's work deserved a B, I gave her a B, and therefore Tina's grade was justified. But I think this is a very strange way of thinking about justification. Tina's grade can be justified, it seems to me, if, and only if, I go through Tina's work, read everything, and make a determination based on my sound judgment. Only under those conditions would Tina's B be justified, even if it had turned out that the randomly given grade was exactly what she deserved.

Understanding the function of grades as a conduit for communicating judgments and not as a carrot or a stick is critically important to understanding what I've just said about omelets and scrambled eggs and how this ultimately relates to the environment and voting.

THE RIGHT AS THE JUSTIFIED

Why a small community of ranchers may establish unwritten rules and codes to govern behavior in commons areas is relatively straightforward: to ensure that there are resources to return to year after year, to make a profit, and to survive. These are all plausible descriptions of their motivations. If any of these is their motivational reason, then it is perhaps a justifi*able* reason as well—that is, it's possible to deploy this reason in the service of justifying the act.

But according to what I've just said about justification, it is not yet (or not necessarily) a justifi*ed* reason. To establish such a reason as justified, it must be subjected to a rigorous justificatory procedure, similar to the one I used to assign grades to my students or to what the courts use to assign a guilty verdict to a criminal. In order to determine whether a motivating reason is justified, we must reexamine our private set of beliefs and desires but also look at our set of embedded commitments, social and otherwise, and subject these to scrutiny. It is conceivable that the conventions that govern ranching behavior are in place for no other reason than that some drunken cowboys have prevailed at darts. But to know this, we must answer the question *why*, and we must do so with an eye toward justification, not explanation.

The famed physicist Richard Feynman once noted in a BBC interview how complicated these explanatory answers to *why* questions can be.[6] Once we have answered the question of why Aunt Minnie slipped on the ice, for instance, there are an enormous number of possible explanatory questions that this answer then raises. Why was Aunt Minnie on the ice? Why did she go there? Why is ice slippery? And so on. Feynman is mostly right about this, though he underestimates the number of answers to *why* questions by half.

When nature strikes, there is only one line of reasoning that can answer a *why* question: explanatory. Mantell, Buckland, Cuvier, Ostrom and others didn't need to go through a lengthy process to justify whether such-and-such an event was right or wrong. It is we humans who interpret the ensuing state of the world as either fortunate or unfortunate, good or bad, and it is we humans who seek to understand what has happened to bring about that state of the world. Observers like Feynman can explain the event by appealing to the principles of physics, chemistry, biology, or one of the other sciences.

In the case of human actions, there are at least two sorts of answers to *why* questions: explanatory and justificatory. We've explored the explanatory at length, and here I've given a brief introduction to the justificatory. In the past several chapters I've raised a string of reasons that one might buy organic peanut butter, geoengineer the climate, or get a flu shot. Some of those reasons appeared to be good reasons, some appeared to be bad reasons, and some were shoulder-shruggers. Could go either way—maybe they're good, maybe they're bad, or maybe they're acceptable but possibly outweighed by other factors and considerations.

It's hard to say for certain whether any particular justification is good, bad, or a shoulder-shrugger, but we often do, and even more often ought to, engage in the project of trying to sift out the good from the bad reasons. I subject each paper I grade to a process of evaluation because this is what is required of me. In order to determine what grade my students should get, I have to go through the process of evaluation. As I do this, I often do so privately, but before I put a mark on the paper I ask myself whether I could justify (and explain) my decision to give the grade I've given. To explain, I would offer my reasoning as a fact about how I came to give that grade. To justify, I would maintain the possibility of reconstructing my reasons and their inferences so as to make a compelling case for the grade.

It is important to the obligations I have assumed as their professor that I justify my grades, that I have good reasons for assigning their grades; but it is also important to me as a person that I maintain and fulfill these obligations. On pain of inconsistency, I couldn't rightly call myself a professor if I didn't go through these motions.

A related but slightly different process must inform our engagement with upselling bartenders, snuggly spouses, and dying grandmothers. Being a son of so-and-so with such-and-such propinquity to my family provides me with binding reasons to visit my sick grandmother. Being a citizen provides me with yet different reasons. How that does or does not conflict with other projects is an open question.

The bindingness of any given norm or commitment is up for debate, of course, and it will to some extent depend on how deeply I've committed myself to a particular identity. Some identities are conditional, some are unconditional, and some are constitutive. Conditional identities, like our friendships and acquaintances, are such that sometimes our commitments can easily be defeated by other reasons. "Sorry I missed our weekly card game last night. I got caught at the office." Some other identities, however, are really quite unconditional. I have responsibilities to my son that I simply cannot break. "Sorry, work, my son needs me."

All of this is well accepted. But there is one ironclad rule of being a person, of being a human animal, of being a creature that takes actions, a creature that does not just create events. When we take actions, our actions are guided by reasons that are endorsed and held by us. These *reasons* are what make us human.

I don't want to deny that we can, at times, abandon our humanity, just as Randall Lee Smith abandoned his. But if being reasonable, if being humane, is important to us—and I suspect that for most of us it is—then the giving and exchanging of reasons is the linchpin identity that ties everything together. If you don't have good reasons for your actions, if you can't defend your actions, if your reasons can't withstand the scrutiny of outside parties, then you might as well abandon any thought of yourself as reasonable.

Philosophers sometimes distinguish between the rational and the reasonable, defining the *rational* narrowly to refer to acts that target self-interested ends and the *reasonable* to refer to acts that appeal to and are guided by fair principles, standards, and values. On the view that conceives of actions as rational, if I can achieve the same results with whiskey and a dartboard, I would be a fool to put my time and energy into reading through all of a student's materials. On the view that evaluates actions as both reasonable and rational, I am not a fool. The elaborate justificatory process of grading is what is required of me. And you are not a fool for voting, for changing your behavior because of climate change, or for having reservations about vaccinating your children.

What, then, is the nature of the principle that sends 130 million Americans to the polls every presidential election season? They don't do it because it's rational. It's not rational. They do it because it's reasonable. It's one of the many commitments that we all take up as citizens in a democracy. So too with some environmental problems. It's not necessarily rational to forgo vegetables grown with pesticides. Organically grown vegetables are expensive. The benefit to you is negligible compared with the cost.

It is, however, reasonable. You ought to act to reduce environmental damage not because your actions will have any noticeable effect but because you have *good reasons* to—reasons that could be acknowledged as valid by any person of sound mind. These reasons include concerns about being a good environmental citizen and, in the case of the producers of food, treating people fairly and respectfully by not trespassing on their well-being by introducing toxic substances into their lives against their will. Now, to be fair, these reasons will certainly also include some consideration of the effect of pesticides on human and nonhuman life—so impacts and outcomes really do matter here—but since the effect of a small amount of pesticide on the vegetables you consume is negligible, this cannot be the only factor. You must also, as with the schmoo shots, respect the willingness of your neighbors to object to these pesticides, and the only way to do this, it seems to me, is to ask them what they think.

Most of the environmental problems we face today, as far as I can tell, are largely a result of the failure to do this; they emerge from our failure to justify our actions. Ignoring for a moment the elaborate defenses of this or that good that my philosophical forebears engage in—and you should read them all: Plato, Aristotle, Bentham, Mill, Rousseau, Kant, Marx, Muir, Leopold, among many others—when politicians and activists offer alleged justifications for their actions, they almost never dig deep enough to get at the underlying implications of the particular doctrine of the good to which they subscribe. Instead, this or that doctrine of the good typically becomes ossified in our political discourse, such that we can map out the landscape like so many moral cartographers, explaining what a political interest group is up to yet entirely ignoring the reasons for what it's doing.

We strive to do right even when the results of our actions will not clearly make a difference. Our "good reason" is that we are dependent on collective cooperation to make important institutions function. You should vote not because you can't achieve the same outcome some other way; you should vote because that is the process by which a democratic presidency becomes legitimate. You should vote not because you expect that your vote will make any aggregate difference, but because if it did come down to a single vote, you would be at a loss to justify your nonparticipation.

So do you want to know why you vote? You vote because you should, not because it's good for you. Voting is what responsible citizens do, like fulfilling promises, telling the truth, snuggling with partners, and visiting Grandma in the hospital. Voting is an obligation built into democratic citizenship, just as snuggling is built into intimate partnerships. A democratic society populated entirely of nonvoting citizens is not a democratic society in any respect. If you endorse the principles of democracy, you endorse the obligation to vote.

Do you want to know why you should take shorter showers, change your lightbulbs, buy recycled toilet paper, and avoid bottled water? You should do all these things, in many but not necessarily all circumstances, because that is what being a good environmental citizen involves: justifying your actions and acting according to those justified reasons. Our mandate as humans, as citizens of the world, is to seek answers to *why* questions that do not play fool to jokers like Sutton. We must ensure, in all that we do, that our actions are justified.

"Out of the crooked timber of humanity, no straight thing was ever made."

—Professor Kant, University of Königsberg

8

The Axes of Evildoers

In which the author (1) turns attention to wrongdoing by invoking the case of the Two Elk fire, (2) shifts the discussion tangentially to consider the first-person attitude, (3) implicates this attitude as a symptom of an orientation toward success, (4) embroils the blogging community by raising the problem of epistemic closure, *(5) reinterprets the principle of charity as benefit-doubt analysis, and (6) suggests that we can reevaluate and reorder our reasons by appealing to the practice of justification.*

FIRE AT TWO ELK

To start a blazing inferno in the highlands of Colorado, one need only a match and the will to cause extraordinary destruction. A bit of gasoline or lighter fluid would help, but the world burns pretty quickly in these parts.

How to do this discreetly? It's simple, really:

1. Identify a suitable target.
2. Gather a plastic milk jug, a sponge, trick birthday candles, and lots and lots of gasoline.
3. Fill said milk jug and douse said sponge with gasoline.
4. Wedge the sponge through the gap in the handle.
5. Insert the unlit candle into the sponge.
6. Light the candle.
7. Run.

You have until the candle burns to its base to get out of Dodge.

In October 1998 several members of the guerrilla environmental group the Earth Liberation Front (ELF) set fire to the Two Elk ski lodge in Vail, Colorado. The next day they issued a menacing missive on the local public radio station: "Putting profits ahead of Colorado's wildlife will not be tolerated." Witnesses reported seeing flames from more than 20 miles away.[1]

Lucky, then, that the highlands did not burn, that the forest did not go up like a tinderbox; that the bears, elk, lynx, and cougars that roam and graze and hunt at 11,000 feet did not scamper frantically from their homes. Lucky that no one was injured, that the costs were a relatively minimal $12 million, that the only really nasty outcome was several chilly ski seasons *sans* chalet at which the 300,000 or so annual visitors might hang their wet socks and warm their frozen toes by the fireplace. Comfort and luxury must sometimes be placed on hold while dedicated crusaders attend to more urgent matters.

The activists responsible for the destruction sought to stop Vail Resorts from developing Blue Sky Basin, though it's never been entirely clear how burning a ski lodge might accomplish this. Ski resorts have nothing if not deep pockets, and removing a building could easily strengthen their resolve to build bigger and higher.

What actually happened was that Vail did build bigger and higher, and the environmental community, rather than rallying around the inferno, began reevaluating its methods. Shortly after the fire, a hot discussion among activists ensued, centering on the utility, and ultimately the futility, of acts of property destruction, offering yet one more reason why my friend Fire—she who had foolishly anointed herself with that incendiary moniker—had made a very bad choice indeed.

What possible reasons could the perpetrators have had for torching the lodge?

1. The environment is precious and must be protected at all costs.
2. Corporations are oppressors of people and must be challenged or stopped.
3. The environmental movement needs leaders and must be promoted.
4. It is cathartic to break the law, and such a thrill must be experienced.

These are all better or worse normative reasons for starting the fires, any of which might help justify the action. Like most answers to *why* questions, however, many of these justifications are faulty and can be shown to be so through rigorous critical assessment. In a somewhat roundabout way, I've been arguing against the first reason throughout this book, but any of these justifications could have derailed on many counts, and the first not just on the line of argument I've been laying out here. Perhaps the beliefs are false. Perhaps the desires are perverse. Perhaps the values are askew.

Here's another set of answers to the same *why* question. Why burn the lodge?

1. To anger the luxury class by destroying its playground
2. To force the corporate powers to capitulate by hitting them where it hurts
3. To get the attention of the press by creating a spectacle
4. To shift the terms of the debate by pushing a radical position
5. To build support for environmental action by showing what a small group of dedicated environmentalists can do.

All of these are arguably better or worse tactical justifications, bound up in the normative reasons I mentioned earlier. These tactical justifications are fundamentally different from the normative justifications that precede them, however. They play second fiddle to the ends that they purportedly support. Each, to some extent, offers a rationale for burning the lodge, and in doing so appeals however tacitly to the first sort of *why* question. Almost any environmental organization you care to name is engaged far more energetically in the tactical discussion than in the wider justificatory discussion.

Tactical justifications are "fact of the world" matters. If you destroy the playground of the rich, you either will or will not anger the luxury class. If you hit the corporate powers where it hurts, they either will or will not capitulate. If you push a radical position, you either will or will not shift the terms of the debate. Answers to these questions are almost always pure speculation, though they can, of course, be informed by political science and sociology.

One problem is that sometimes people do very bad things—like, exceptionally bad things—in the name of doing very good things. Another problem, perhaps the underlying reason that people do such bad things in the name of doing very good things, is that it is hard to distinguish authentic normative reasons from brute reasons if we turn our attention primarily to the tactical questions. Radical environmentalists engage in these actions, and others engage in related political actions, because they believe themselves to have "right on their side," effectively jumping the justificatory gun. Whether an action is right (or morally justified) is completely buried by the question of whether an action will work (and is justified on tactical grounds).

Many might hear this as a sort of "the end justifies the means" reasoning. We saw in chapter 6 that the end alone can't justify the means. We can't, for instance, simply look to better states of the world—a more stable climate, a schmoo-free world—to know what to do. We also have to look at the reasons an action was taken and run them through a justificatory procedure—a test, of sorts, to ensure that they are reasonable, rational, and impartial. Again, this isn't to say that sometimes a very good end can't override other ethical considerations. It's just that the question of whether any given action is permissible should not be assessed by looking strictly at the end. The means, as well as the reasons for the action and the responsibilities of the agent, also have to be taken into consideration. Nevertheless, when most philosophers say that the end justifies the means, they mean that *if* we have identified the proper end, *then* we can rest assured that any means we use to get to that end are justified as well.

Five days after the terrorist attacks of September 11, 2001, President George W. Bush announced that his administration would rid the world of evildoers. "Tomorrow," he said, "when you get back to work, work hard like you always have. But we've been warned. We've been warned there are evil people in this world." He continued, "My administration has a job to do, and we're going to do it. We will rid the world of the evildoers."[2] Bush's comments were immediately seized upon by the commentariat for their simplistic black and white depiction of good versus evil.

Within six weeks, on October 26, at the beginning of a prolonged case of capitulatory legislative dyspepsia—dyspepsia spurred by fear of future terroristic reprisals and resulting in all-too-frequent imperialistic eructations, including the invasion of not one but two countries, the systematic abrogation of international agreements prohibiting torture, and the gradual encroachment of constitutionally protected freedoms, among other international embarrassments—the United States Congress passed the noxious Patriot Act, which included language not only addressing increased security measures, but that explicitly roped environmental activism into the concept of terrorism, thereby making incidents of property destruction punishable as crimes against the state. Several of these attempts at lawmaking sought to lump even nonviolent environmental civil disobedience with ecoterrorism.[3]

To buttress its case on the Two Elk fire, the FBI opportunistically convened a special counterterrorism task force, christened it with another menacing militaristic name Operation Backfire —the operation naming division of law enforcement must hover breathlessly over the trees of Hollywood hills, lying in wait of the next fantasy blockbuster, hoping beyond hope that their catchy operation names will have scooped the shoot-em-ups—and accused the Vail arsonists of ecoterrorism.[4]

Disregarding the thorny question of what, ultimately, terrorism is and, more obscurely, what distinguishes ecoterrorism from other terrorism, we can nevertheless state with certainty that the eco-saboteurs considered themselves to be engaged in an altogether nonterroristic, eco or otherwise, action. According to their own statements and missives, they took every precaution to ensure that nobody would be hurt in the burning of the lodge.

In chapter 6 I closely examined a decision to take an action by asking whether we could apply a more robust account of reasons to our forward-looking decisions, specifically regarding geoengineering and the schmoo vaccination. It became apparent there that the work of identifying the right decision simply could not rely strictly on the outcome. In chapter 7 I proposed that we actually make decisions all the time independently of whether the outcome will be good or bad. I used the example of voting to make this point. I hoped to show also that many environmental decisions, such as how to change our behavior regarding information about climate change, are also of this nature. Now, however, I want to look at justification in yet a slightly different light. I want to address the so-called *publicity of reasons* to briefly explain that the reasons we have—motivational or justificatory—aren't simply reasons but are always already subject to the downward pressure of outside evaluators.

WHO'S ON FIRST? WHAT'S ON THIRD?

One hundred and ninety-two years after Mary Anning unearthed the bones of her paradigm-shattering dragon, but a mere five years before Vail's treasured Two Elk lodge went up in flames, the video-game *Doom* blasted onto the burgeoning computer-gaming scene. It was one of the first graphics-rich multiplayer video games that could be played simultaneously at locations around the world. Nine game levels were distributed by shareware, and within two years of its release more than 10 million players had been infected with the zombie bug.

Together with a collection of other controversial pixelated hell-raisers—including *Mortal Kombat, Resident Evil, Modern Warfare, Call of Duty, Grand Theft Auto, Medal of Honor*, and eventually *Left Behind—Doom* was responsible for kicking off the craze in a variety of video games called *first-person shooters*. In the game, you (aka "*Doom*-guy"), a space marine stuck on Phobos (one of Mars's two moons), must battle zombies and demons as they spill from behind blocks and walls into your line of fire. Like most video games, it's a ridiculous fiction, but kids are drawn like moths to all kinds of ridiculous fictions, particularly if those fictions flash and beep, so the game took off like a Colorado wildfire.

Doom and games like it have been criticized for, among other things: their graphic violence, excessive gore, nudity, sexism, dehumanization of others, exaltation of criminality, invocation of witchcraft, surrealism, unrealistic imagery, and, yes, even their high-processor demands.[5] You name it, someone's criticized it. The games have almost all been charged with inspiring a spate of real-world violence, including numerous homicides and suicides, among which the school shooting at Columbine High School in Littleton,

Colorado, stands out as the highest-profile. For a time rumors swirled that Eric Harris and Dylan Klebold, the two teenage shooters, had wiled away their preshooting hours meticulously creating a level of *Doom*. When children act violently, if they have had any exposure to violent video games, fingers start pointing.

For the most part, I believe, this criticism is soporifically simplistic. Video games corrupt impressionable youth by imprinting on the blank tablet of the mind images/experiences/realities/ideas that these otherwise innocent children might not be exposed to. *Yes, we know*. Violent video games lead kids to emulate the violent behaviors of the characters that they play. *Yep, got it*. Players grow so desensitized to the concerns of others that they cease to treat people in the appropriate way. *Blah, blah, blah*. The criticisms are well enough known.

Of course, violence is not unique to video games, and defenders of games are quick to point this out. Movies, television, radio, rap music, heavy metal, punk rock, blues, jazz, theater, Dungeons & Dragons, and so on are replete with death and dying, murder and mystery. Even Socrates was known to offer a few nasty words against the poets, all the while bearing the brunt of the criticism as a corrupter of youth.

One might, however, argue just the opposite: that the first-person point of view—which is unique to this variety of video game—gives the players a personal vantage that allows them to crawl into the twisted minds of the wretched characters they play. Indeed, one could further reason that this first-person perspective would therefore be good at instilling empathy in the player. The novelist Nicholson Baker has noted that games like *Call of Duty* sometimes exhibit a kind of Janus-faced schizophrenia. When a player dies in *Call of Duty*, for instance, the game closes with a palliative aphorism—a quote from Gandhi or Confucius, for instance—about the benefits of peace and the perils of war.[6] Clint Hocking, creative director of the game *Far Cry 2*—a game applauded for addressing the moral dimensions of war by responding to the continued violence of the players in such a way that it eventually becomes evident to the players that they are the cause of, rather than the solution to, greater violence—points out that games like

Badge of Honor and *Call of Duty* have nothing to do with honor or duty.[7] It's all very curious, given the heavy invocation of morality and virtue throughout.

Despite the vantage that first-person shooters give to players, the games don't really contribute much to the first-person experience. The players' only sense of what it might be like to be "*Doom*-guy" facing an onslaught of zombies is visual and audial. What it is actually like to be wandering through a dangerous labyrinth remains entirely obscure to the players. It's true that the players *view* the world through another's eyes, but they bring to the video screen a raft of real human experiences and social presumptions that forever cloud any pretense of what it must be like to go on a zombie-killing rampage. To suggest that this somehow gives the players the "experience" of being the character is like suggesting that because I've driven your car, I now understand what it is to be you. Ironically, the first-person standpoint can be just as objectifying as the third-person standpoint.

Whether images of violence cause, give rise to, or plant the seeds of such violence, we will probably never be able to say. Some empirical studies have backed up these claims.[8] Others, not so much. More compelling is that some studies have concluded that violence, however depicted, needn't even be serious or graphic; repeated actions oriented toward causing intentional harm to another are enough to trigger violence in children.[9] In short, it ain't the violence in the game, it's the way the actions in the game reorient the player.

There is a goal in Doom: to get to the end of the game, to survive, and to mercilessly kill and destroy every moving object, living or otherwise, that comes along. Players are to treat all enemies as forces of nature, will-less nonagents, even if they appear in the form of human beings. If a player aims to win the game—and honestly now, what player doesn't aim to win the game?—the rules permit, and in fact encourage, annihilation or obstruction by any means necessary.

In the case of *Doom*, the player is out to kill zombies. I'm no expert on zombies, but my repeat viewings of *Night of the Living Dead* as a reluctant tenth-grader, sheepishly clutching the corner of my sleeping bag, lead me to believe that these monsters are hideous, lecherous, stinking, rotting, drooling corpses, motivated by a singular horrifying objective: to suck out your brains. They thrive on taking life from nonzombies and using them for their own purposes. More important, their consumption of nonzombie brains is their *sole* objective. They have no other.

Like the monsters and cryptids we discussed in chapter 3, zombies share some characteristics with humans (they are, by definition, human beings devoid of humanity), but their singular motivating drive (to kill nonzombies and eat their brains) is not dissimilar from the many other motivating drives that one might locate in nature's beasts. (Of course, many critters in nature's bestiary have social drives to care for their young and form communities, and even, in some cases, to be gentle and kind. What many animals lack, however—except for marginal cases like dolphins, dogs, mountain gorillas—is the capacity to justify their actions.) Zombies are therefore full of motivating reasons but devoid of normative reasons. There is no reasoning with these creatures. They act only with some degraded form of rationality. In essence, they are brutes, endowed with human ingenuity and bodies but unable to parse through arguments. They instead maintain the killing capacity of dumb machines, reinforced by the inhumane drive to do so.

In many depictions, as in *The Walking Dead*, zombies are also pack animals, attacking in waves: torrents of dehumanized human beings who act together to fulfill their objective. In this respect, they're not only easy to understand, they're also easy to exterminate. It's very easy to kill zombies because they don't give any lip. They don't argue, they don't challenge, they don't even ask politely: "Um, excuse me, I see that you are trying to prevent me from eating your brains. Would you mind refraining?" Instead they just explode, or fall over, or collapse the instant that their brains are pierced with a spike.

And that's somewhat odd. One might imagine that I am raising this point to illustrate how similar to a zombie a killer like Randall Lee Smith is. But that's really only part of what I'm getting at. I think we must also acknowledge that for killers like Smith, it appears to have been relatively easy to murder Susan Ramsay and Robert Mountford Jr. Why? For the same reason that it is easy to blast zombies in *Doom*: there is no uptake, no point at which one can say, "Hold on a second. Let's see if we can talk this out." The pleas that Ramsay and Mountford might have offered in their own defense—"No, please don't, I don't want to die"—simply did not resonate with Smith. He was not in a position to measure his reasons against their reasons.

Doom-style first-person shooters illustrate with great poignancy the problem with keeping our own heads while assuming the roles of others. On the one hand, it is clear that the hordelike zombies are driven toward a singular end. On the other hand, the shooter appears authorized to obliterate the zombie because the zombie is driven solely by this end as well. It's a two-way street. An orientation toward success enables this systematic, dual-pronged nullification of others. It permits the pushing around of people like objects, like pawns on a chessboard.

This problem is not straightforwardly about dehumanization, as one might initially assume. I agree that games such as *Doom* dehumanize enemies, effectively transforming them into objects. But most games we play, arguably *all* games we play, dehumanize our opponents. To explain this dehumanizing feature away, we might draw the conclusion, as many before us have done, that it is therefore the graphic violence or the degree of violence that is problematic. I draw another conclusion: it is the games we play, and how we relate those games to other ethical contexts—not, in other words, the content of those games, but the nature of games and the way in which they familiarize us with a certain class of rules—that leaves us scrambling for clarity on the rules and constraints to which we are subjected. It is no accident that the characters in HBO's serial *The Wire* routinely refer to their direct or tangential involvement in the drug trade as "the Game."

None of this is to say that games themselves are problematic. I love games. But games, and the rules that guide them, are applicable only in a very narrow context. We are mistaken if we think we can extrapolate wider claims about moral rules and reasons from them. The question, therefore, is not whether the games are so vivid and realistic that they lead the vulnerable to mimic them, but what is involved in compellingly playing these games and whether this translates appropriately into a nongame environment.

THE ROOTS OF ALL EVIL

About 10 years after the events that reduced the Two Elk lodge to embers, the FBI's case against the phantom arsonists broke. The perpetrators were discovered to be members of an organization out of Eugene, Oregon, that referred to itself as the Family. Ten activists were involved in coordinating the attacks, and the Vail arson was hardly the first in the Family's criminal career. All told, some 20 firebombings in Oregon, Washington, California, Wyoming, and Colorado have been blamed on the Family.

Arrested and charged in connection with the Vail fire were Bill Rodgers, Chelsea Dawn Gerlach, Stanislas Meyerhoff, Kevin Tubbs, and Jacob Ferguson. Two others, Rebecca J. Rubin and Josephine Sunshine Overaker, fled immediately after these arrests, but Rubin turned herself in to the police in 2012. As of this writing, Overaker is still on the lam. Their comments and fates after the arrests are telling.

Rodgers, the alleged ring leader, suffocated himself with a plastic bag in a Flagstaff jail cell just before he was to be shipped off for trial. Gerlach was accused of 18 counts in five attacks. She claimed in court to be motivated by "a deep sense of despair and anger at the deteriorating state of the global environment." She continued, "I realized years ago that it was not an effective or an appropriate way to effect positive change," adding that "the firebombings did more harm than good."[11] Her regret over the bombings she was involved in included remorse over her poor tactical decision to firebomb the lodge. Meyerhoff pleaded guilty to 54 counts in seven separate attacks. Rubin, who had been fleeing police for some time, was sentenced to 5 years in prison, 200 hours of community service, plus $14 million in restitution. The lesser perpetrators got plea bargains.

Tubbs has written the following in a letter posted on a website created to support him:

> Regarding the actions I was involved in, I'd like to point out that I was only motivated by a passion to save animals and a desire to help the environment so that we could pass on a healthier and a more vibrant Earth to our children and to future generations.

> In these actions, I always acted out of love—never hate or malice—and I was spurred on and motivated by the continual and enormous scale of ecological destruction taking place as well as the extreme cruelty and disregard for all animals and life forms that is so prevalent today.
>
> I helped out with these actions only with altruistic intentions. I had nothing to gain personally by helping out; not money, not glory, not fame. Not eternity in paradise with 72 virgins.[12]

What is perhaps most striking about Tubbs's reasoning is that it invokes the terrorism of 9/11, and the alleged justifications of terrorists, to try to plead on his own behalf. If you recall, for quite some time after 9/11, the preposterous idea circulated that the 9/11 hijackers had committed their horrible crimes because they earnestly believed that they would be rewarded in the afterlife with their own planet populated by 72 virgins. The problem here is that this logic—so simple and clean on one hand, yet so ludicrous on the other—is actually an explanation, not a justification. It *explains* the hijackers' behavior. It doesn't *justify* their behavior.

This explanation, of course, is way too easy. For one thing, it reduces the hijackers to mindless zombies. Criminals and murderers they certainly were. Zombies? Far less likely. For another thing, if the explanation is wrong, the prescription for a solution will also likely be wrong. If the hijackers were under the illusion that they would be greeted by virgins in the afterlife, then this is a problem with their beliefs, and perhaps best addressed by attending to these beliefs. If instead they were acting out of a commitment to oppose Israel's occupation of Palestine, then a wholly different response would be called for.

The point here is that reverting to explanatory interpretations reduces the justificatory reasoning of the actors into a set of possibly conflicting motivational explanations, simply explaining their behavior in motivational terms and doing very little to engage the full span of justificatory reasons (both normative and brute reasons) that may ultimately be guiding them.

Again, it's important to point out that, in these cases in particular, their justificatory reasons need not be particularly strong. Upon reflection, any reasonable and rational evaluator can easily say that there are no "good" reasons that could possibly justify their behavior. (Indeed, this is the view that almost all people hold regarding 9/11 and similar attacks.) The point is that by mistaking explanatory reasons (the reasons that explain the behavior) for justificatory reasons (the reasons that would justify the behavior), we miss the moral import, the ethical upshot, of their actions, making it difficult to blame them for wrongdoing. How, after all, do you blame a zombie? In so doing, we foreclose on the appropriate response. This isn't a problem with their action per se, but a problem with whether they are motivated by, and can be said to act according to, good or bad normative reasons, better or worse justifications.

(It is worth noting here that the scholarly literature on explanation and justification is extensive, confusing, and extremely technical. There are also disputes among philosophers about how best to deploy these distinctions. Some people argue that justificatory reasons should be kept entirely separate from motivational reasons, whereas others argue that justificatory reasons can be included in the private set of motivations of individual actors. I don't want to get bogged down in this discussion. I'm giving here only a kind of cartoonish overview of the distinction between explanation and justification, smoothing out many of the difficult spots, because I think the distinction can be incredibly helpful when assessing actions. This will undoubtedly upset some of my colleagues. If you're really interested in the technical discussion, there is much more to read.)

Consider grades again. When I give my students their grades, I do my best to ensure that each grade is justified. What I mean by this is that, if called upon to justify a grade—say by an angry or confused student—I could offer up reasons that would lend force to my conclusions about the merits of their work. There's one area, however, where I might run seriously afoul. Some students are very good at their coursework but are nevertheless not very strong students; they perform well on assignments, but they do not, on closer examination, really understand what's going on. They might, for instance, be conversant in the material and give the impression that they know what's going on but nevertheless completely misunderstand it. In these instances, I might give them a grade that I believe to be justified, but nevertheless only be able to defend that grade only on the basis of their work and not their comprehension. This is, in fact, one of the ways in which students can seek to "game the system."

Students can take one of two orientations toward a class: they can seek to get an A in the class, come what may understanding-wise; or they can seek to gain a grasp of the material, come what may grade-wise. Most students fall somewhere between these orientations and aim to understand the material but also do whatever they can to get as high a grade as possible. The first prioritizes success over understanding; the second prioritizes understanding over success. Ideally the two purposes are in sync, so that students who have learned and understood the material also do well in the course, but this is not necessarily always the case.

In turn, a norm of behavior emerges from the priority of one over the other. The students should understand the material. They should be taking my class, or going to the university, for the right reasons. The right reasons in a classroom are those associated with the orientation toward understanding. The wrong reasons are those associated with the orientation toward success and getting the grade.

There are certainly ways to game the system. In many classes one can simply memorize the lecture notes. If you become an efficient memorizer of lecture notes, you're bound to get an A in the class; this is particularly true for courses that emphasize learning rote facts, conveying information, or transmitting knowledge. More advanced courses certainly test a student's comprehension of the material. Math classes may be a bit like this because the student is asked to think through and solve problems. In these cases a student can do one of two things: learn the operations and unreflectively chug through them, or understand the operations and reflectively process them.

The same can be said of many activities: games, politics, work, family, friendships. There are at least two ways of orienting oneself toward these areas of life: either with an orientation toward winning, or with an orientation toward doing them right.

Games like *Doom* illustrate one type of possible orientation, because the players aren't playing the game properly if they don't play to win. The orientation toward success requires not that we think about the perspective of another person but that we think seriously about how to get an opponent to do what we want—essentially permitting the transformation of all things, human and nonhuman, into objects and obstacles.

We can easily filter any of our decisions through this strategic lens. Need to clean the kitchen? Put the kids in front of the boob tube and throw those dishes in the washer! Need to get to work on time? Clear the roads, bad drivers be damned! Need to meet a profit target? Streamline the company and pink-slip the low performers! It's very easy to flip on the virtual reality helmet of the orientation toward success and treat people as though they are pieces to be moved around on a chessboard. In these instances, it's not like violent images or video games have enabled actors to thoughtlessly transform others into objects or tools or even servants. It's an attitudinal thing, brought about by the very human drive to get things done and accomplish stuff. In order to accomplish stuff, we need to orient ourselves toward our goals, and sometimes this involves viewing the world as a set of obstacles and aids. This attitude is made all the easier because, of course, it is true that people are, among other things, *objects*, as are animals and trees. And it is also true that we can sometimes use people to do our bidding. The world is filled with people who will get in our way or help us achieve our goals. A store clerk may prevent us from walking out the door without paying for a product we really want. A mover may help us pack up our belongings and move across the country.

Granted, these are voluntary activities on the part of the store clerk and the mover (caveats about being "forced to work in order to feed one's family" aside), but they are also some of the many ways in which we routinely adopt an attitude toward people that views people as means to an end. Yet this is only really an attitude, an orientation, a perspective that we adopt to bend the world and the people in it to help us get what we want. If you want to get something done, it would behoove you to understand how people work. It helps to steer them. This is why we see our political pollsters and marketing professionals conducting extensive studies of human behavior: to better understand what drives people to certain behaviors. In this way, however, it is an attitude that militates against justification in favor of understanding the world in terms of explanation: how best to achieve our ends.

More expansively, then, one can also take an orientation toward doing the right thing. We don't have as many everyday, ready-made examples of situations in which we ought to take an orientation toward doing the right thing. Just as a student's burden is to choose the orientation toward understanding, which is the right attitude for a student, the burden of any person is to orient oneself toward justification, toward a position in which one's actions have gone through the requisite process of justification.

The important point isn't that there are two ways of viewing things but that one way is fundamentally deficient but nevertheless extraordinarily prominent. It is an after-the-fact orientation that threatens to cut out the justificatory enterprise that has heretofore characterized morality, putting the tactical cart before the justificatory horse—making, in effect, moral action obsolete. Through the operational lens of success, practical and tactical action engulfs moral action. Money is sometimes said to be the root of all evil. But the root of all evil is not money itself, it's the *pursuit* of money, the pursuit of success.

EPISTEMIC CLOSURE

It is common nowadays to drive all political discussions into the gas chambers and crematoria of Nazi Germany. Godwin's Law, a tongue-in-cheek restatement of the laws of the Internet, states, "As an online discussion grows longer, the probability of a comparison involving Nazis or Hitler approaches one."[13] The state of political discourse in the United States has certainly reached the level of the absurd. The health-care debate of 2009–2010 illustrated this rather neatly, as did the McCain-Obama presidential race of 2008, in which fallacies of association were used to great effect by the opponents of Barack Obama.

Godwin's Law makes a strong showing in the environmental arena as well. References to Hitler and the Third Reich are open to all comers: Glenn Beck has made absurd charges about Al Gore, IPCC Chair Rajendra Pachauri about Bjorn Lomborg, Senator James Inhofe about all believers in global warming, NASA climate scientist James Hansen about all coal companies, Viscount Lord Christopher Monckton of Brenchley about young climate activists, and so on.[14]

Vilification of one's opponents is easy enough. Also easy enough is the old "Am not! Are too!" routine, insisting that one person has the right justified positions whereas the other does not. It is easy to get caught in a loop. In the spring of 2010 conservative blogger Julian Sanchez made waves in the Internet community by christening this loop "epistemic closure," suggesting that the structure of current political discourse is such that ideological pressures force media outlets and commentators to shut out critical input from opposing platforms.[15] (The term *epistemic* comes from *epistēmē*, the Greek word for "knowledge," and thus means "of or relating to knowledge or the acquisition of knowledge." There is another use of "epistemic closure" in theoretical circles that technically describes a very different phenomenon related to knowledge. When Sanchez appropriated the term to use in a political context, quite a few philosophers protested.) Republicans get their news from Fox, Fox gets its news from the right-wing blogosophere, the right-wing blogosphere gets its comments and views from various Republicans, all of which in turn influences the Republican Party. Similarly, Democrats get their news from MSNBC, MSNBC gets its news from the left-wing blogosophere, the left-wing blogosphere from its democratic readership, all of which in turn influences the Democratic Party.

Sanchez was criticized for suggesting that epistemic closure is more a phenomenon of the Right than the Left, but that was the silly side of the conversation. The more interesting observation is that epistemic closure happens at all and appears to be relatively common. In fact, recent research into cultural cognition and psychology suggests that this kind of epistemic tribalism is fairly common throughout our political discourse.[16] The lesson, of course, is that we would all—regardless of our place on the political spectrum—do well to guard against it.

The problem here is that epistemic closure of this variety ends up stanching alternative views. Well beyond a mere echo-chamber effect, epistemic closure sets the justificatory discussion to the side and transforms political and moral discourse into tactical reasoning requiring explanatory answers. Suppose, for instance, that because I am a member of a particular epistemic community—say, the community of environmentalists—I believe myself to be justified in taking some action X to protect the earth. If I am a member of this community, the question of what I *should* do, or what I *ought to* do, becomes less about what my goals ought to be and more about how these goals can be accomplished. In this way, justifications within a closed epistemic loop very often come in the form of operations: if you want to accomplish X, you must do Y. The imperatives are hypothetical, contingent on the objectives. There is no leakage, no external input about the rightness or wrongness of the objectives. There is no reason to doubt that X is worth doing. Such epistemic closure thereby masks the *why* of doing X with the *how* of doing X, by turning imperatives and prescriptions into a complex mapping problem.

Considered together, then, closed arguments can generate an action that may be quite wrong, not necessarily because the end is bad, but because the means taken to get to that end cannot be justified by that end. So, back to Vail: Maybe it would be great to stop the expansion of ski resorts in the Rocky Mountains. That's a view I can accept. I live here in the Rockies, and I'd like to keep our ski industry from trampling nature. But even if I think that I should keep the ski industry from expanding, "justifying" the tactic of burning down the lodge is much harder. (It is curious and important here that *justification* takes on a double meaning. What we actually seek to do in this case is not to justify the rightness of the act but rather to show that the act will be successful. This isn't really a justification at all; rather, it's a kind of explanation.)

Sure, you might quibble with the arsonists' math. The arsonists were mistaken in thinking that they would productively affect

Vail Resorts's bottom line. Vail has deep pockets. So they miscalculated. But that, again, is really just a tactical calculation. It can only do the work of "justifying" the act—that is, explaining how it might work—within a very constrained set of considerations. In no small part this is because such explanatory arguments crowd out and disregard the variety of extremely important objections held by those who may be affected by X but nevertheless sit outside the justificatory community.

Closure is a function of the orientation toward success, toward winning. Once we've allegedly done the hard work of justification, we're free to close our justificatory loop and aim at our goal. So if we've identified that we should steer the world in the direction of fewer endangered species, or a lower population, or a cooler climate, then we need only think about how best to achieve this end. The conditions of play are set, the rules established, and the players are unleashed on one another to kick and hit and scratch and bite until a conclusion is reached—all in the name of pursuing this goal—without once stopping to reevaluate the stance that setting the rules according to the goal is an upside-down and inside-out way of thinking about things.

In political discourse, this is sometimes characterized as pursuing a special interest, as though all interested parties are up to the same sorts of shenanigans. But special interests come in at least two flavors: private and public. Private interest is asserted by individuals, organizations, and companies driven by their own financial interests or personal well-being. When Monsanto injects its concerns over DDT into the public discourse, obfuscating the arguments of Rachel Carson, it is acting in the pursuit of its private interest. It acts monotonically to advance its bottom line. The public interest, in contrast, is asserted by individuals, organizations, and companies acting to advance some special issue that only may or may not have something to do with the private interests of the individual actors who make up the organization, but probably is making an appeal to the greater public good. Ducks Unlimited, the Conservation Fund, the Sierra Club, the Nature Conservancy, and Earth First! are all special interests, organized around the general unifying principle of a particular aspect of the environment.

Advocates for these and other related groups frequently seek to demonstrate why the interests of the group are actually also in the public interest.

What both special interests share, however, is this orientation toward success, toward winning, toward advancing their narrowly or widely prescribed good. They are *political* groups. This may explain why many people confuse the two types of special interests. Your beliefs about the justifiability of a particular project—whether the project be the growth of one's bottom line or the promotion of a specific cause—will determine whether special interests are abrasive to you.

This objection to special interests comes up again and again in our political discourse. It is an outrage, for some, that private interests have been able to exert so much influence over our political process. Critics cry foul over undue influence in the political process. Advocates argue that there are no interests better suited to participate in the political process than heavily invested stakeholders. For many others, however, it is not the private interests that are out of bounds, but the public interests. When environmental, labor, or women's organizations take to the streets to advance a position that is aligned with their stated interests, *this* is the outrage, to the critics, who charge that public interests are pushing for a privileged position for their pet projects. Advocates point out that special interests exist to promote values that ought to apply to all.

To take a step back here, orienting ourselves toward success is, under many circumstances, perfectly reasonable. We want to throw a good (successful) dinner party, so we head to the store and make our decision about chicken or fish. All we need to know are the likes and dislikes of our guests. (I mean, yes, we need to know other things, like how to cook and how to entertain and how to be cordial, but we don't need to know much more than this.) As we have seen, much of the time the moral constraints are already in place. We rely on others—legislators, regulators, inspectors—to ensure that our food is just and safe. If we want to throw a good (justified) dinner party, however, we need to do more. We need to justify our actions. This involves identifying the constraints ourselves: Ought we to consume mahi-mahi? Is there anything morally problematic with eating chicken? Is there any reason for us to be concerned that we are lavishing ourselves with extravagant foods while there are poor and hungry people just outside our door? (Note also that just as there are dual senses of the term *justification*, so too are there dual senses of the term *good*.)

What we need to do is to find a way of checking our claims and actions against themselves. We can do this by keeping the justificatory process open. An open justification, in contrast to a closed justification, doesn't offer the same sort of operational directive.

BENEFIT-DOUBT ANALYSIS

Opening the justificatory loop is no simple task, but it can be accomplished by continually revisiting one's reasons for taking action in the first place, by taking one's motivating reasons and testing them for justificatory strength. This is perhaps best accomplished by opening moral deliberation to a range of parties: shifting from an attitude of Doom-like first-person dominance to an attitude of second-person interaction, laying bare our reasons, and ensuring that in full sunlight they make sense. In effect, opening motivational reasons to the scrutiny of others flips the usual transaction on its head. Instead of first justifying an action and then motivating ourselves to act according to those allegedly good justifications, opening the justificatory loop involves identifying the underlying reasons motivating the action, subjecting them to outside scrutiny, and ensuring that these reasons are justified.

As a feature of this, morality is a two-way rather than a one-way street. It is incumbent upon everyone to seek out the reasons for their actions and scrutinize them for plausibility and justifiability. If someone doesn't support immigration out of a belief that it harms the economy, it is important to test the justifiability of this view: not just whether it may or may not harm the economy, but whether harm to the economy is an acceptable reason not to support immigration. Maybe it's true that one should support immigration even if it *is* harmful to the economy. The way to do this is to create channels and institutions through which epistemic loops cannot become closed and through which justifications are routinely tested by others. Just as one does not want to release a book to the public that has not been vetted by external referees, so too ought one not to allow one's justifications for action to go untested.

What's wrong in many circumstances is that we seek to justify our actions according to some presumed good, whatever it may be—sometimes, counterintuitively, that good takes the form of adherence to principle or duty or commandment—and we do so in a way that places priority on the good over the right. I suspect that we are all, to some extent, wrong about any particular conception of the good. I mentioned in the last chapter that we carry our reason-rich identities around with us, that these are wrapped up in various commitments to various conceptions of the good. Naturally, these tend to be ideas that we endorse. It is easy to fall into the trap of thinking that our conception of the good is the only correct conception of the good. Presumably we believe and endorse these identities for a reason. What else are we to do than to stick by our beliefs and defend them?

Here's the deal. Sometimes we make mistakes—big mistakes. We accidentally wave to a stranger, thinking the stranger is one of our friends. We calculate our tip incorrectly and upset a chatty waitress. We misjudge a person's character based on our distaste for tattoos. We make a decision on a purchase without having fully considered the implications of that purchase. These mistakes occur in no small part because our knowledge of the world, and our epistemic situation, is narrow, limited, and to some extent closed. If our emphasis on doing the right thing depends on our conception of the good, then no matter how careful we are, we are doomed forever to pile mistake on top of mistake. If, however, we conceive of the right as the justified, where justification is an ongoing checking-in process oriented toward "doing it right" instead of winning, then I think we can avoid epistemic closure and avoid the mistakes of evildoers past.

This means that moral justification is a fluid two-way street, involving not just an obligation on the part of actors to get their rules right but also an obligation of nonactors to ensure that others get the rules right too. It means, essentially, prying open closed justificatory loops if they are oriented too narrowly toward winning, which can really be accomplished only if both actor and nonactor, speaker and hearer, adopt an attitude toward doing it right.

A fundamental principle that I teach my students is to give the proponents of positions a charitable read. In the courses I teach, we call this the Principle of Charity. Generally, I advise extending this charity as far as possible, even to reprehensible people with ugly objectives and disgusting personal habits. If some political opponent claims to support a position for a reason that one finds stupid or inane, it is wise to seek to understand the position at face value, maybe even to give the person the benefit of the doubt and find the strongest argument for him or her. We teach this in philosophy because we want to spend our time assessing the strongest, not the weakest, arguments. This stands in sharp contrast to the orientation toward winning arguments that is taught in some rhetoric, law, and politics courses.

Consider: The objectives of the Family may have been well-intentioned. They wanted to save the earth, and the expansion of the Vail complex seemed to be a perfect target. Did they act for the right reasons? I seriously doubt it. They acted for the wrong reasons. They were deeply wrong about their reasons, even though they had the best of intentions.

OVER THE RIVER: REORDERING OUR REASONS

If a child expresses excitement about a trip to his grandmother's house because she gives him gifts, this is cute for a time. But as the child matures into adolescence, it becomes less and less cute. In these cases, it is not just a violation of etiquette that the child expresses these sentiments. It's an indication that the child is excited for the wrong reasons.

What I've been doing in this chapter is setting the stage to argue a point related to the priority of the right over the good, arguing similarly that justification has priority over application. Suppose I want to get to grandmother's house. How should I go? That's one question. Over the river and through the woods.

Why am I going to grandmother's house? That's an equally important yet different question, and we must ask it of all who travel to grandmother's house. The *how* I go is caught up in the *why* I go.

Sometimes my reasons are pure: I go to grandmother's house to see Grandma, because she is an important person in my life. But sometimes my reasons are less than pure: I go to grandmother's house because my mother forces me to go, or because I hope to be showered with gifts. Nothing about my action has changed in either case: I have gone to my grandmother's house. But there's something distinct about the case in which I go to my grandmother's house for the gifts versus the case in which I go out of an interest in seeing my grandmother. If my grandmother finds out that I went to see her because my mother forced me to go or only because I hoped for gifts, I suspect that neither my grandmother nor I would feel very good about this. So too if others learn that I am the type of person who feigns interest in my grandmother only for the gifts she gives me. Like many, however, I'm pretty good about keeping my motivations to myself, so I can usually get away with having ugly motivations on one hand, but pretending to have good motivations on the other.

I think the only way to make sense of this is to chalk it up to a distinction in types of *reasons*. In one case, my reason for visiting my grandmother emanates from a set of normative commitments that I share with her and the rest of my family. In the other case, my reason for visiting my grandmother emanates more narrowly from my brute, subjective desires. The question is whether the reasons that motivate me are of the normative sort, meaning that they are, have been, and/or could be defended against the scrutiny of other observers, or whether they are of the brutish sort, meaning that they have not been, and possibly could not be, defended if they are exposed. The former is, effectively, a case in which my will is involved in the decision making. It's the case in which I can be said to have done right or even brought about good. The latter is, effectively, a case in which my wants and desires usurp my decision-making authority, in which I can be said to have done wrong by my grandmother, and even brought about bad.

The same can be said for the Vail arson. If the activists' reasons for burning the lodge were to collect insurance money or to exact revenge on Vail Resorts, these are goals that would be very difficult to justify. If, by contrast, their reasons for burning the lodge were to save the forest, then this only may or may not be a laudable or morally praiseworthy goal. Saving the forests, of course, is probably a praiseworthy goal. What will determine its moral praiseworthiness, however, is not only whether the forests must be saved, but whether the forests are saved in the right way. In order to determine this, the justification for the act, along with the means of acting, must be subjected to the scrutiny of outside parties.

For many people, the destruction of the Two Elk lodge was an act of terrorism, different only in scale from the actions of the 9/11 terrorists. Whether you agree or disagree with this characterization of the two events, it is easy to see that the two events at least share the features of being (1) more than *mere* events, and thereby (2) actions that demonstrate hardened resolve on the part of their perpetrators—a resolve that is partly reinforced by a commitment to a conception of the good.

This all raises this important observation about *priority*. Many philosophers debate about the priority of either the right or the good. Those who defend the "priority of the good over the right" seem to think that if we can just figure out what the good is—happiness, health, money, welfare, nature, freedom, Elvis—then our practical directives will fall into place. Those who defend the "priority of the right over the good" suggest just the opposite. They suggest that our practical directives do not so clearly fall into place. Our practical directives remain subservient to concerns about the right.

At the beginning of the chapter I gave two lists of reasons, each in response to the same *why* question: Why burn the lodge? The first list presented primarily ethical and normative reasons. The second list present primarily practical and descriptive reasons. What very often happens in our political discussion is that the two sets of reasons become intertwined, so that the practical/descriptive reasons begin to look and smell like the ethical/normative reasons, or in some cases completely take the role of ethical deliberation.

Consequently, many of our attempts at achieving the good confuse ethical justification with tactical justification, effectively supplanting normative reasons with descriptive claims, explanatory statements of fact or speculation about what will work. In lieu of normative justifications about what we ought to do, we bicker over tactics, "justifying" our actions by appeal to predictive models of human, animal, or natural behavior. These models can get quite complicated, *but they still miss the point.*

For instance, it is sometimes claimed that if we aim to halt resource depletion, then we are justified in limiting the population and therefore justified in limiting individual families or rationing in a certain way. Once we establish what the best state of the world should be, it is thought, we will have our answer about what is justified. It's a simple matter of choosing the most expedient course of action. But this is wrongheaded, as I've been trying to show.

I've been trying to say that we must instead engage, in an ongoing and open way, the project of justifying our actions.

• • •

The annual UN Framework Convention on Climate Change meetings that I mentioned earlier—in Berlin, Geneva, Kyoto (birthing the Kyoto Protocol), Buenos Aires, the Hague, Bonn, Montreal, Bali, Poznan, Copenhagen, Cancún, Durban, Paris to name most but not all of them—offer a case in point. These meetings are widely viewed by many experts inside and outside the environmental community as a monstrous and abject failure. The meetings rarely yield the results that environmental advocates hope for, and often no conclusion is reached at all.

To my mind, however, they function really very well. There are 20,000 or more people at these events, all working together to focus on solving a collective problem. Much emerges out of these meetings, even if no collective agreement emerges: coalitions, philanthropic bodies, activist alliances, international partnerships, information sharing, deliberations, disputation—all furthering the purpose of addressing climate change by seeking to reconcile the concerns of multiple parties.

In another respect these meetings certainly *are* an abject failure. If viewed as ends-oriented events in which stated objectives were not achieved, then they've failed miserably. We have not yet resolved the problem of climate change. But the "problem" of climate change is a multiheaded hydra. It's never been entirely clear what the fundamental *problem* is. Is it that the climate is changing? That we're changing it? That some are more responsible for changing it than others? That it will adversely affect us? That it will adversely affect others? That it will adversely affect some more than others? That it will affect our ecosystems?

The question here of what one should do, of whether an action was effective or appropriate, illustrates the underlying tension in this book. Whether an action is effective is not a morally robust answer to the question of what we ought to do.

The orientation toward doing it right, toward justification, is a deeply democratic orientation, guided by the intuition that there are mutually acceptable means—that is, non–war mongering means—for us to sort everything out, for us to achieve some mutually acceptable, though not necessarily mutually beneficial, resolution of our differences. This is the basic idea that guides a lot of my work as well as the work of other philosophers, ranging from Immanuel Kant to John Rawls to Jürgen Habermas: that we ought to be able to (and thus can) abstract from our particularized predicaments long enough to determine what is right.

Orienting oneself toward success essentially enables a primitivistic drive to win, to beat, to dominate sense and reason out of the debate. It is supported by evolutionary and naturalistic thinking about our actions. It is reproduced in many of our descriptive models of human action.

Politicians who argue for a policy based on the number of votes that it will win them, companies that greenwash to garner customers, scientists who embellish findings to communicate urgency, skeptics who throw every manner of obstacle in the path of environmental policy to prevent adoption—this orientation toward success is the underlying nature of politicization. It involves putting policies into place for the wrong reasons.

At the same time, it ought not to be understood as *mere* politicization. Just as we can ask *why* questions about the Family's reasons for burning the Two Elk lodge and offer Willie-Sutton-style explanations to these questions—they did it because they are jobless hooligans looking to call attention to their cause—we do so at the risk of dramatically oversimplifying their reasons and completely misunderstanding their aims and objectives. If instead we seek an answer to the *why* of the ski resort attacks by appeal to justifications—they did it because they had a grievance, legitimate or illegitimate, and despite the costs they saw this as the most effective way of addressing it—we are in a much better position to unravel the enigma of their misdeeds.

So our question really is considerably more complicated than it initially was made out to be. How do we preserve our humanity, how essentially can we be human, while also acknowledging that we live in a world of finite resources and that our activities here may shape and affect the lives of others over there?

I've suggested that the most responsibly green position we can offer is the position that seeks justification for an action, that asks at every decision juncture whether an action can be justified, that seeks justification before application. Contrary to what many environmentalists think, the wrong approach for our environmental woes is the approach that suggests adopting an attitude toward strategy and winning, toward preserving nature. Being ethical and acting rightly is not, and cannot be, about winning some sort of geopolitical game. It is not about bringing about an environmental paradise in which everyone drives hybrids and children's cribs are festooned with ivy. Rather, environmentalism must be about having the right reasons. The way in which we know that we have identified the right reasons is by ensuring that our reasons are regularly subjected to the scrutiny of a wide community of diverse evaluators.

9

The Green's Gambit

In which the author (1) addresses positions that underdetermine the role of reasons in the establishment of value, (2) picks on arguments aiming to establish ecosystem services as identifying the value of nature, (3) offers the reader a price on her kidney, (4) shifts the discussion from value to rules, (5) introduces the first and (6) the second formulations of Kant's Categorical Imperative, and (7) suggests that rules and laws expand autonomy and freedom.

DON'T FEED THE BEARS

Anyone who has ever gone camping knows that it is laughably stupid to feed the bears. They know that leaving food out at night, intentionally or unintentionally, may present a problem later on and that dripping s'mores on one's pajamas or spitting toothpaste by one's tent is perhaps not the best way to ensure a restful sleep. They probably even understand that their safety is partly determined by campers who have come before them and that their own nonstupidity is essential for the safety of the campers who will come after them. If they are unlucky enough to set up camp at a site that has been frequented by sloppy campers, they may well encounter a bear even if they themselves have been diligent about keeping a clean site.

There are, of course, less selfish reasons to be concerned about feeding bears—ecological reasons, for instance. The bears, we are told, will grow dependent on our food offerings. They will cease to be self-sufficient. A common bumper sticker in camper country reads "Garbage Kills Bears," stating unclearly whether the bears' taste for human food will keep them searching for more or the bear populations will grow so large and aggressive that we humans will be forced to pay for our sins by culling them.

Curiously, bear logic infects our attitudes about an extraordinary range of nonbear issues. We also ought not to feed the poor, some people say, because the poor will cease to be self-sufficient and will start relying on us to feed them. Lieutenant Governor Andre Bauer of South Carolina was excoriated in January 2010 for making comments along this line. "My grandmother was not a highly educated woman," he said, "but she told me as a small child to quit feeding stray animals. You know why? Because they breed. You're facilitating the problem if you give an animal or a person ample food supply. They will reproduce, especially ones that don't think too much further than that. And so what you've got to do is you've got to curtail that type of behavior. They don't know any better."[1] Garrett Hardin made a related, albeit less offensive, case in his famous and much maligned article, "Lifeboat Ethics: The Case against Helping the Poor." His reasoning was that we ought not to feed the poor because the earth cannot sustain a voracious population.[2] Outside of prudence—the poor are not likely to sneak into our campsites at night—the thinking is similar.

If the asininity of this comparison has not yet clobbered you over the head, perhaps I can spell it out more succinctly. "Poor people" are not bears. They are people. True, they may be people whose actions and behaviors can be understood by appeal to predictive principles that describe the movement of families and populations, but they are also people who can respond to reasons. We do not know the poor's reasons for being poor. We do not know their reasons for living the lives that they live. To talk of "the poor" as a giant festering mass of generalizable individuals is to disregard any given member of this category, and yet our political discourse is rife with this sort of bear logic.

Hardin was roundly lambasted for his piece when it came out in 1974, and he is still to some extent read as a foil today. Few others deign to take such a stance. But Hardin's approach to environmental problems is nevertheless prevalent in many other ways, and even if not explicitly stated, the prevailing assumption in many environmental policies is similar: we must control the lives and lifestyles of others, in some important human way, lest we face imminent environmental catastrophe.

Not only this, but we open up a pathway for criticism of a most devastating sort. You've heard the objections. "Regulations kill jobs! Moonbat policy makers promote death panels! Garbage kills bears!" At the very least, the caricature that environmental laws run at odds with economic growth is the single most prevalent argument against environmentalism. Both types of argument—for and against heavy-handed intervention—cut in a wide array of directions, many of which appear to fall back on hardened economic principles. But my suspicion is that often the underlying reasoning isn't so much economic as social, and when the contrast arises it is because those who object do so because bear logic surreptitiously sneaks in. Let me explain.

When my students propose, as they inevitably do in the face of the evidence for climate change, or resource depletion, or desertification, or urban sprawl, or simply an uncontrolled population explosion—as Paul Ehrlich did in 1968, whose aforementioned book *The Population Bomb* carried the ominous subheading *Population Control or Race to Oblivion?* and as Ted Turner proposed more recently[3]—in short, when my students propose that the best and most efficient way to approach our environmental crisis is to place restrictions on how many children a family can have, or to seriously constrain the consumption of others, or to completely and utterly stop consumption of some important resource, they fall prey to the same intuition that has guided so many environmentalists before them. They supplant humanity with nature.

Heavy-handed policies that neglect the core concerns of people—about family, health, safety, security, freedoms, and rights—in the name of preserving nature or becoming more sustainable are not just unjustified, they are *in principle* unjustifi*able*. They could not, if put to a test of public scrutiny, be justified to those affected. They treat our environmental issues primarily as problems to be solved, and in so doing do not open those problems to the solutions of the people whom they will affect. It is in this respect that the solutions are heavy-handed: they stealthily deploy assumptions about the good and the bad without questioning the extent to which these issues are properly characterized as problems or issues or crises with characteristically closed solutions.

It's not that we don't have serious environmental problems and issues and crises; and it's not that we can't or ought not ever to place heavy-handed restrictions on ourselves and/or on others. It's that our tendency is to think of environmental problems/issues/crises operationally instead of justificatorily. The science of economics, the statistical methods of sociology, the probing of psychology, the physics of physiology, all point in the direction of a position that views human behavior in terms of the best explanation, that offers resolution to problems in terms of tactics and strategy. We isolate a problem, we identify possible mechanisms by which the problem can be shunted, and we take action to shunt the problem. It's clean and neat—except that it's not clean and neat, morally speaking.

I don't mean to disparage the importance of empirical tools. They are deeply helpful in our understanding of the workings of humans and of the world more generally. But they are only tangentially related to the justificatory endeavor, subservient to the overarching justificatory enterprise; which is, in essence, a *human* enterprise—maybe even *the* human enterprise. Justification, I have been trying to show, can be accomplished only by bringing people together, actively engaging in the project of justifying our actions, and getting them to sort out and evaluate the variety of reasons that knit our moral and social fabric together—making the eggs instead of simply making egg dishes.

On the standard way of viewing things, environmental problems are critically important because it matters that the earth maintain a stable climate or that a species be spared the ignominy of extinction. On this way of looking at things, the way to show that the environment matters is by arguing that a stable climate or a species's survival is important or valuable or good. If we specify some climate (C) as good, then we have a problem if, and only if, the climate deviates from C, which it does and will. We can then introduce a bunch of interventions to bring the climate back to C.

But on a totally different way of thinking about things, climate change or species loss is a problem—*our* problem—mostly because it is we who are creating the changes, because we are unjustified in bringing about these changes. It is this discussion about justification that we, living citizens of the world—mere hatch marks on the grand intergenerational timeline—are not having. Or if we are having it, our discussion is situated primarily in the language of costs and benefits. What I've been trying to suggest is that we are justified in bringing about an altered climate only if we have actually done the hard work of justifying our actions. In the case of climate change, by virtue of the grand-scale tragedy of the commons, I think it's fairly clear that we have not done this hard work, and we're certainly making very few efforts to try to justify making this change. We don't need to cap our population or stifle our economy or step on the backs of the destitute just yet. What we really need to do is to reevaluate the ground rules.

In the previous chapter I suggested that the environmentally right action is the one that is taken for the right reasons. In this chapter I want to clarify some criteria that might be used to distinguish the right from the wrong reasons.

HE WHO HAS THE GOLD MAKES THE RULES

The gold standard in the environmental economics literature, and to some extent in the environmental policy literature, has for many years been cost-benefit analysis (CBA), or, somewhat differently, cost-effectiveness analysis (CEA). Almost every policy that emerges from legislatures is run over with a fine-toothed comb regarding the cost of the policy and the benefits it will confer, whether in initiation or in execution. The Endangered Species Act (ESA) includes provisions that insist upon the employment of cost assessment once a species has been listed. The National Environmental Policy Act (NEPA) includes provisions permitting cost assessment to be used either monetarily or nonmonetarily in an environmental impact assessment.[4] The Clean Air Act (CAA) and the Clean Water Act (CWA) do the same. Consideration of costs and benefits is also a key feature throughout public deliberation over the robustness and continued utility of the acts. Despite the cost-assessment requirements of all of these environmental policies, some policy makers and industry groups have taken legal action to integrate CBA more centrally in the establishment of environmental policy than it already is.[5] So it will be helpful for us to more closely examine the differences between the two types of analysis.

Cost-benefit analysis requires that before imposing or implementing a restriction or a protection, we weigh alternative policies: Is the proposed policy worth it? Do the benefits outweigh the cost? In contrast, *cost-effectiveness analysis* requires that in order to implement a restriction or a protection, we weigh alternative implementation options: What's the most efficient way of implementing this rule? Though they seem to be similar, the two are actually importantly different, and these differences have given rise to heated debate over the years. In essence, this is a debate about *moral priority*: whether the good has priority over the right or the right has priority over the good.

The appeal of CBA is self-evident. It offers access to the alleged right course of action by aggregating costs and benefits, essentially steering policy in the direction of conferring the greatest overall benefit to society in general. To put this somewhat coarsely, CBA proposes that the *right* thing is what is *best* overall. The good has priority over the right. Who wouldn't want to create overall benefit at a reasonable price? But we have already seen in the geoengineering discussion that simply improving the lot of the world isn't enough to authorize intervention. I can't paint and repair your house without your permission, even if it will improve your position considerably. I also ought not to make a determination about whether a species is endangered and deserving of protection only if the benefits outweigh the cost. There are strong arguments against prioritizing the good over the right.

CEA, in contrast, proposes that the *best* course of action is that which most efficiently implements the *right* rule. In this case, the right has priority over the good. Many standing environmental regulations reflect this concern and place a different sort of constraint on our actions, imposing rules that aid in guiding cost-benefit analysis but do not place priority on it. These rules aim instead to steer the justificatory process, to clear the way for cost-effectiveness analysis to answer the complicated question of how *best* to do the *right* thing. That's a big difference. The costs and benefits of any given action are still important considerations, but such considerations, most environmental policies propose, should come into play only after prior constraints on our behaviors have been factored in.

The problem, in part, is that is a good bit less intuitive than CBA, which carries incredibly persuasive weight. The self-evident nature of CBA gives the politician a stance to bring to the floor, a persuasive argument to present to the public. CBA aims at maximizing benefits and minimizing costs. Who wouldn't love that? But CBA also gets the moral ordering exactly backward. We ought not to do the right thing only if the benefits outweigh the costs. We ought to do the right thing first and foremost; then, once we know what we ought to do, we should do so in the most cost-effective way possible.

Moral priority is reflected directly in the procedural priority of the policies. The rules as currently written aim first to establish the reasonableness of moving forward with an endangered or threatened listing, and only then to look at the potential costs and benefits of doing so. This only makes sense. The idea, naturally, is that in order to know whether we should protect a species, we must first know whether it is endangered. Doing a cost-benefit analysis before a determination of threat is made puts the cart before the horse. It implies that only those species that are valuable should be considered for listing. It's like asking whether you can afford to buy orphan meat before you consider whether orphan meat ought to be on store shelves in the first place. Not only is this a bad way to make dinner plans, but there are also an infinite number of things that you might be able to afford to buy that you otherwise shouldn't.

Many in the environmental movement are rightly skeptical of cost-benefit analysis, and they argue vehemently against it. Objections to its widespread integration pepper the literature. They tend to express the following concerns: (1) that cost-benefit analyses are ineffective, giving only lip service to protection of the environment, (2) that methods of cost and benefit approximation are problematic, deficient, or incomplete in some way, (3) that they rely on incommensurable comparative methodologies, (4) that they are too burdensome or onerous and therefore a waste of resources, (5) that it creates political gridlock, and so on. As compelling as these criticisms of CBA are, there's a kind of tacit acknowledgment in the environmental community that CBA is just how policy is done. Perhaps not unreasonably. The quantitative and comparative measure it offers is extremely valuable to politicians and policy makers alike.

The problem is that these criticisms are aimed more at *refining* the method rather than arguing against it. What is ignored, unfortunately, is the role of moral priority. What would instead be helpful is an alternative approach to justifying environmental policies that establishes the moral priority of the right over the good.

ICE BATH

Travelers to foreign countries will be familiar with the urban legend of the tourist who, after a night on the town, wakes up in a bath of ice next to a sign that warns him not to move. One of his kidneys, the sign explains, has been removed.

What price would you place on your kidney? How much would you be willing to accept in exchange for your kidney? On its face, this is an innocent question. Suppose you answer that you would accept $10,000 for your kidney.

> **Exchange:** I am willing to pay $10,000 for your kidney, and you are willing to sell your kidney for $10,000.

In the economics literature, this is known as *willingness to pay*. If we make the exchange, all is copacetic. There are lots of things that economists think we can measure in terms of willingness to pay: apples, oranges, pets, plots of land, ecosystem services, personal health, and so on.

But now suppose I ask you again how much you'd sell your kidney for, and you again answer $10,000.

Steal: One night while you are sleeping, I enter your room, leave $10,000 on your dresser, and take your kidney. You wake in an ice bath to discover that you are without a kidney but are now $10,000 richer.

On one way of thinking, this is the right kind of exchange. On another way of thinking, there's something deeply troubling about this arrangement. A reason has been disregarded—a reason specific to you: I neglected to get your consent.

It would be a mistake, however, to construe this merely as a problem of consent. Economists have clever devices for getting around this. Maybe, for instance, we place a monetary value on your consent. Suppose, for the sake of argument, that this price is exactly $5,000. This raises another possibility:

Consent Buyout: By running complicated economic studies of your past behavior, I calculate the value of your consent at $5,000. As a result, one night while you are sleeping, I enter your room, leave $16,000 on your dresser, throwing in an extra $1,000 for good measure, and take your kidney. I have taken your kidney, but I have compensated you both for your kidney and your consent.

If the view that values consent is correct, my compensation regime should present no moral problem. It should be exactly equivalent to the state of affairs in which I compensate you for your kidney and receive your consent. But I think it's fairly easy to see that there is still an objection to my doing so. The objection is that it would be much, much better not simply to compensate you for your consent, but instead to acquire your consent (through the usual means) and then to take your kidney. What I'm effectively doing in the above case is *buying you out* of your consent, and this is not something that many of us will countenance, at least not without putting up a fight. Consent is not the sort of thing to which a price easily sticks.

Those who object to this position like to point out that there is *some* price, maybe a very large price, that would make such a move acceptable to you. I'm sure that's true. $10 million left on your nightstand and you'd probably feel pretty good about your sudden fortune. Nevertheless, there is a remainder, a lingering sense that something has gone unfulfilled, that the comparative universe in which your consent is secured and $9,995,000 is deposited in your bank account is a much, much better universe than Steal. You can change the numbers to suit, but no price on your consent will overtake or flush out this remainder.

Here's at least one reason that you might find this arrangement problematic: you have no information about the use to which your kidney will eventually be put. In most cases of commodity sale, this needn't bother you. If someone buys your old jogging suit, you relinquish control of the suit, and all the uses to which the new owner puts the jogging suit are more or less justified by virtue of the purchase, everything else being equal. He can wear the suit to work out or to a Halloween party. He can burn it or wipe the floor with it. What was once your property is now his property, and the range of uses will vary depending on the property rights regime to which the new owner subscribes.

Kidneys don't seem to work like that. The range of acceptable uses is far narrower. There are uses to which they *should* be put (to save an ailing child), uses to which they shouldn't be put (to feed the family dog), and uses to which they should disputably be put (to aid an elderly alcoholic).

Moreover, the use will probably affect your consent-buyout price. Maybe you learn that the kidney will save the life of a sick child, so your consent may be relatively easy to secure. If you later learn that your kidney has gone to a kidney collector who intends to keep it in a jar on his desk, you may think this a wretched and unacceptable use for a kidney, and price your consent considerably higher. Your consent-buyout price will shift depending on how the kidney will be used.

Your willingness to part with your kidney is (or at least ought to be) a reason associated with your will. Ironically, the economic concept of willingness to pay flies in the face of the will even while purporting to take the will seriously—what you are *willing* to pay is not equivalent with the payment that can be extracted without the participation of your will. Yet willingness to pay is routinely used in cost-benefit analysis to justify the abrogation of the wills of others. There is a reason for this: according to many economic models, the will is merely an aggregation of values, not a repository of reasons.

A central objection to the commodification of useful organs—or to the commodification of nature, as I understand it—has to do with the apparent incongruency between value and price. It seems unproblematic for us to place some object that has some use value—say, a hammer—on a shelf and offer it up for sale or exchange. This commodifies the object; it transforms it from simply an object with some use, that is valuable for some reason (not necessarily, but often, a use-related reason), to an object with a fixed price (whatever the reason). It is essentially to strip the object of reasons related to its value and winnow those reasons into a singular monetary unit.

An object's use value, presumably, is caught up in the reasons for which it is to be put to use or that people value it. A hammer has use value because it is good for hammering nails, a cup has use value because it is good for holding water, an ax for chopping wood, an opiate for relieving pain. These are all acceptable uses, and we can countenance almost any use for most objects. Hammers can be used to stop doors, cups for bailing canoes, axes for film props, and opiates for getting high.

Not all objects with value have use value, mind you. A sentimental keepsake, for instance, may be valuable for some reasons specific to the owner. Those who've offered the "precious vase" argument for the preservation of nature often seek to demonstrate the nonuse or intrinsic value of nature.

In some cases, as with opiates, we have external laws that regulate our use of these objects so that they are not put to bad or undesirable purposes. Very dangerous medicines, for instance, we regulate extensively. These require a prescription that is supported by a strong medical justification. Tapeworm got you down? Take two of these and call me in the morning.

We also regulate the use of objects by constraining the actions for which they may be used. We prohibit murder, for instance, but

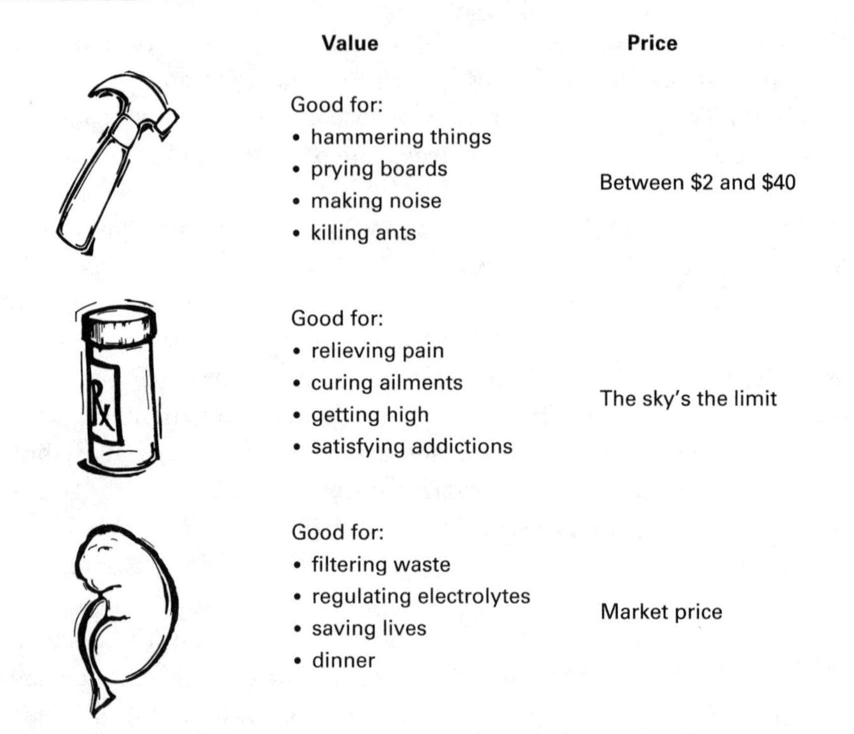

we would be mistaken if we restricted the use of hammers and axes simply because they are used in some murders. In most commodity arrangements, it ought not to matter to us what a given object is being used for. So these things are unproblematically characterized as commodities. We permit the commodification of an object, which thereby shifts the reasons establishing its use value over to a unit of price. Other objects, like guns, present a much more controversial problem. They are great for hunting and for forming militias but also particularly effective for killing. We may need to have a wider discussion about the reasons that one is purchasing a gun. We don't need that discussion for tanks and fighter jets.

The conventions of exchange in the free market, and in some cases the rules and laws of the free market, are such that the parties to the exchange—shopkeepers and the like—are mostly not in the business of interrogating customers about their intents and purposes. If a teenager walks into a store to purchase a hammer, the shopkeeper naturally sells the hammer without needing to know the purpose for its use, without raising a question about the buyer's reason. This is an ironclad rule of free commerce. I can buy cabbages to feed my family, or I can buy cabbages to throw them off of buildings, but I suspect that most of us would agree that a shopkeeper would be out of line to query me on my reasons for purchasing cabbages. Even if there are some circumstances in which it's reasonable to ask why I'm buying cabbages—perhaps the shopkeeper knows me and is just curious—she's out of line to restrict me from purchasing cabbages just because she doesn't like my reason, and she would certainly be out of line to charge me a higher price because she thinks that I have bad reasons.

These rules of free commerce are made possible because of the "thing-like" qualities of cabbages and kidneys. By stripping use- and other value-related reasons away from the thing in question, we essentially commodify it, transforming it into an object with mere exchange value. Such commodification isn't a huge problem for objects like cabbages, but it gets very sticky when we begin talking about objects like our bodily organs, which we presumably control and maintain some authority over our reasons for using. Most of us would presumably like to maintain this authority, or at least regulate the uses to which such objects can be put.

Unfortunately for nature, and for beings that live and frolic in nature, the constellation of animals, vegetables, and minerals that surround humans are all bona fide objects—rocks, mountains, streams, water, air, etc.—and thus relatively easily commodifiable. Being a commodity—being "thingified"—carries some pretty awful downsides. True, some of the things that surround us are living objects (trees, shrubs, lichens, fungi), and some of the things that surround us are autonomic objects (parasites, bacteria, viruses, genetic material)—but they are, mostly or wholly, objects. If they are not objects, they are often instead events made up of many objects acting in symphony: earthquakes, tornadoes, rainstorms, hurricanes, volcanic eruptions. To date we've done a decent job of preserving humanity and not over-objectifying other people—though even here there is certainly a trail of wreckage—but as reasons-respecting creatures, we must also pay attention to other value reasons that are too easily reduced to a price. The way to do this is not by redirecting the discussion toward these varieties of value but by preserving the reasons that underwrite the value, which can be done placing *priority* on the right over the good.

THE RULES OF THE ROAD

The dispute over the priority of rules stems, in part, from a misapprehension about both the nature of rules and our understanding of rules. As with many of the topics I've been raising briefly, there's a robust discussion in philosophy about this, and my explanation below should not be taken as the last word.

Some people believe, for instance, that all rules, moral rules included, are conditional, meaning that they're binding only by virtue of whether they will bring about the desired ends. To be sure, many rules *are* conditional. Suppose you want to rejuvenate the cod fisheries in Newfoundland, which collapsed in 1992 as a result of overfishing. One might then suggest that it is imperative for fishermen to immediately cease fishing in these waters. In this case, it is *imperative* that they do this, but there are conditions on this imperative. If we don't *want* to return the cod fisheries to their original state, then all bets are off. The fishermen needn't stop fishing. Many imperatives are like this. Do we, or do we not, want to revive the cod fisheries?

Others believe, however, that some rules—moral rules, in particular—are also unconditional and apply in all circumstances, regardless of my wants or of the balance of costs over benefits. I ought not to kill innocent fishermen, for instance, regardless of whether I aim to rejuvenate the cod fisheries. In this case, the rule that one ought not to kill is *categorical*, or unconditional, since it applies in all circumstances. Even if my objective is to restore the fisheries, I am not permitted to kill in order to restore those fisheries.

That's easy enough to understand. But trace this out a bit.

Many conditional rules are contingent on our wants and objectives, much like the cod fishery. But many other conditional rules are governed by the logic internal to them, like the logic of cooperation, say, where we find rules of pure coordination, such as those dictating the side of the street on which we drive. Our decision about the side is contingent—could be left, could be right—but it is the *logic* of cooperation that keeps the rule functioning. If we coordinate, we'll all get to our destination faster. These rules are binding not just because we have made them so but because they are of the sort that must be followed in order to function. If there is a side of the street on which we must drive, then everyone must drive on that side of the street; otherwise the rule won't work. Since road safety is presumably something all (or many) of us want, a world without traffic laws would be a bad world.

Many unconditional rules are like this, too. Rules, like those insisting that we must always keep our promises, that we never lie, that we repay our debts, or that we act for good reason, are not so much rules of pure coordination, but they *are* rules to which we are bound because of their internal logic. The rule that one should always keep one's promises, for instance, follows directly from the idea of a promise, which is an obligation under which one places oneself. On the one hand, it is a construction, but on the other hand, it's an incredibly important construction. It creates an environment in which individuals can count on others to do what they say they will do. A world with promises is a world in which we are free to rely on one another. It's a world in which we can place ourselves under self-legislated commitments to be fulfilled at some future time in some future place. And it is equally a world in which one can request that others also place themselves under binding commitments. When these rules are subscribed to, they create the conditions under which we can consider ourselves to be rational and reasonable. This creates the conditions that make further choices possible. It is tempting to construe such rules as also contingent and conditional, as also stemming from our wants and desires, but such a stance casts aside the implications of a universe in which there are no promises. In a universe without promises, in which one can never know if a commitment will be fulfilled or abandoned, we would hardly be recognizable as rational and reasonable, and we would forever mistrust one another.

Most rules fall somewhere in between the conditional and the unconditional, but still bind us not by our wants but by their internal logic. Modern banking institutions work like this, as does our insurance industry. Such institutions are not only essential financial tools, thereby governed by conditional regulations not unlike those that tell us which side of the road to drive on, they also enable our freedom in much the same way that promising does, and they are thereby governed by unconditional rules stemming from their logic.

The rule that one should always pay one's debts, codified in law but grounded in morality, for instance, enables us to take out loans on our future. If we don't have money now, we can instead borrow money from a lender, and we should and must pay this debt, because not doing so will undermine this very important institution. This rule is deeply important to us. It expands our freedom greatly. It is easy to think that the reason you should always pay your debts is that the law will come after you if you don't. This may be true. But this is a conditional reason, tied to what you want, not to the underlying unconditional rule that you should pay your debts. Debt is *enabling*—it enables us to do things that make our lives better without having to go through the extraordinary difficulty of pushing a wheelbarrow of money around. In this way, the rule that you should pay your debts promotes the good by requiring respect of the right. The enabling feature of debt is what makes it the right thing to do, not its relationship to how it makes your life better.

All of this then ropes right back around to the super-duper, double-secret, underlying principle of humanity that we discussed in chapter 5. If you recall, the principle states that people, unlike animals, act according to reasons; reasons that are, in principle, justifiable. It is thus an underlying presupposition of all actions that these actions could, in principle, be justified. Insofar as this is an underlying principle of actions (that they are taken for reasons that are in principle justifiable), then people operate according to this principle when they undertake to act. More importantly, if reasons are, in principle, justifiable, and if those who take action do so with this principle underlying the actions they take—if, in other words, they tacitly endorse this principle (which they do, and they must, because they've taken the actions and not simply allowed the events to occur)—then these reasons can withstand the justificatory scrutiny that we discussed in chapters 7 and 8. This requirement, this rule, emerges by dint of the *logic* of the super-duper, double-secret, underlying principle of humanity, like many other noncontingent rules. Thus if one betrays this rule when acting, one is both deploying the rule and disregarding the rule, which is, all told, a sort of contradiction—at least in practice.

All of which is not to say that we can't disobey these rules. Of course we can disobey them. We do so all the time. But we should at least recognize the practical contradiction in doing so; and if we do recognize this, then it seems that we should be rationally persuaded that we ought not to disregard these rules. If we instead seek to temper the application of these rules on grounds that their implementation is too costly, or that the benefit will not be great enough, or that they conflict with other important principles and rules, this is one thing. It is an important strength of our decision-making apparatus to evaluate and act upon the betterness or worseness of outcomes. If, however, we never get to the point where we have the justificatory discussion about which rules, following from which institutions, we are to endorse, because we have already closed off this line of inquiry by dividing the world into goods and bads, benefits and costs, prices and dollar values, then this is totally different.

If, to put this metaphorically, you see a child drowning in the ocean, and you know that you should save him but you come to the considered judgment that it is too risky for you to save him, I think there are reasons to excuse you for not saving him. This is not a duty any reasonable person would hold you to. If, however, you prematurely assess that anyone drowning in the ocean will be too risky to rescue, and so you do not even look to see if anyone is drowning when someone may well be, then I think you are culpable for having neglected your responsibilities. "Swim at your own risk" signs will only excuse so much. If you are standing there, beside the sign, watching someone drown, you are responsible for having allowed this drowning. If called upon to defend yourself to a community of your peers, you would invariably come up short.

The same applies to environmental rules. You ought not to kill an animal that is an endangered species (whenever it suits you); but the reason you ought not to kill an endangered species isn't because the rule is good for you (or won't be bad for you), or even because it will be bad for that individual animal (or won't be good for it), but because the animal is an endangered species. Killing it will likely place the species in further peril. If you were to kill it just for fun, say, and you had knowledge of the endangeredness of this animal, many would charge that your actions were *indefensible*. If called upon to defend yourself to a community of your peers, you would invariably come up short, just as with the drowning child.

Witness, for instance, the fairly regular drumbeat of public objection to the trophy hunting of endangered animals. In July 2015 the social media universe went crazy with protest against Walter Palmer, an Indiana dentist who paid $50,000 for the opportunity to kill a prized lion named Cecil. He was excoriated online and on television by politicians, comedians, and commentators for an act that many people deemed unreasonable and selfish.

It is one thing to kill an individual member of a species, independently of its species relation, but quite another to contribute to the eradication of an entire genetic line from the face of the planet. In the Cecil case, the endangeredness of the species is doing a lot of the heavy lifting. Palmer ought not to have killed an animal that is endangered, not because killing that individual animal might be bad for it, but because that individual belongs to a species that is endangered (and highly valued).

The case of Cecil the lion is complicated by the social importance and charisma of lions, so those sorts of objections might be raised as well. They are also important. Palmer's actions were, and ought to have been, subject to public scrutiny by reasonable and rational evaluators who could, given some distance from Palmer's personal desires, weigh his actions against independent standards and evaluate the weight of reasons that he might have had. What ensued in the aftermath of the initial outcry was a fairly important national discussion of the role of trophy hunting in species conservation.[6]

THE SENSITIVITY OF CAR ALARMS

When television entertainment personality Glenn Beck—clown to many but inspiration to some—invoked in his 2003 memoir the exploits of Hollywood actor and director Orson Welles, proclaiming hyperbolically the profound transformative effect that Welles had had on him, he likely did not realize that he was offering perhaps the best example of illogical and immoral behavior that any philosophy professor could ask for. According to Beck, Welles made it his personal policy to drive through town blaring his horn so that people would get out of the way, acting as though his car were an ambulance or a police vehicle.[7] It doesn't take much reflection to see that if every driver were to make this her own personal policy, it would create a problem for actual ambulances and actual police vehicles. Nobody would ever move out of the way.

There are rules against this sort of thing, laws even, that reserve blaring horns and sirens for emergency situations. Cops aren't supposed to use their siren if they need to get to the grocery store, for instance. Are these mere regulations, like those coordinating traffic or driving on one side of the road?

I think not. I think there's something more to them. As far as I can tell, these sorts of rules stem from logic internal to horns and sirens—established by the reason for these warning noisemakers. Because it is important for us to know when there is danger or when others need help, we need a reliable mechanism to alert us. Sirens on ambulances and police cars offer this aid. Our laws regulate this sort of thing pretty heavily, but it's not hard to see how many of the institutions on which we depend would unravel if it became *de rigueur* to make oneself an exception to the rule. If *everyone* decided to be an exception to the rule, the rule (or the policy or the practice), would simply cease to have any meaning.

For instance, in the early 1990s a good number of fancy cars came equipped with car alarms. Evidently they're still a thing, but there is very little evidence that they deter crime. As most remember, the early car alarms were incredibly sensitive. Walk too close to the car and the alarm would go off. As a consequence, the car alarm became meaningless. It no longer alerted others to a possible theft but was instead merely an annoyance, meaningless background noise. It wasn't simply that everyone was becoming desensitized to the sound of a car alarm; people could still *hear* the alarm, after all. It was that the use of the alarm was inadvertently being universalized. The alarm lost all meaning. It lost its reason for being.

The philosopher Immanuel Kant thinks that this sort of universalization procedure—that is, the procedure of imagining what would happen to a rule if *everyone in the moral universe* were to follow it—can help us get a clearer sense of moral claims. Anybody who has ever taken an introductory philosophy course will recognize the Kantian Categorical Imperative. Here it is in its first formulation.

The Categorical Imperative, Variation I:

> Act only according to that maxim by which you could, at the same time, will it to be a universal law.

The idea, in principle, is that some rules, our moral rules, are so fundamental to our functioning as rational agents, as free wills, that they cannot be abrogated without undercutting rational agency. The Categorical Imperative is meant to act as a road map in helping us understand and unearth these rules, which aren't written anywhere but are implicit in rationality. Kant asks us to imagine a universe in which the rule that guides our action—suppose we wonder whether it is permissible to get to our destination faster by blaring our horn—were, through our taking the action, to be converted into a universal law governing all behavior. Rules that show themselves to be logically or practically inconsistent cannot be correct. A universal law permitting all to use their car horns or an emergency siren to get through traffic would be self-undermining. This is not permitted.

The emergency sirens rule isn't terribly important to rationality, contingent as it is on our need to address emergencies. It depends how important a functioning transportation infrastructure is to us. But other practices are *vital* to rationality.

Rules that permit lying whenever it suits us, for instance, couldn't easily be made into a universally acceptable law, since we'd never know whether the people with whom we were speaking were lying or telling the truth. The boy who cried wolf wasn't merely being impertinent by crying wolf; his tale illustrates the hard lesson of universalizing the norm of crying wolf when it suits us. Because he cried wolf whenever he felt so inclined, the crying of wolf became meaningless to the villagers who could help him. Likewise, the institution of truth telling depends for its strength on the underlying supposition that people will not lie even when it suits them.

For all Kant's fascinating contributions, his moral program has been criticized by many, not least because it is strongly idealistic, but also because it is so formalistic that it downplays the specifics of our unique and particular predicaments. Some circumstances of lying seem plausibly permissible—maybe even required. If a murderer comes to your door demanding entry and clearly states his intent to kill your friend who is hiding in the other room, it would seem that morality demands, at minimum, that you lie to the murderer to get him to go away. If we're serious about morality, claim the critics, we need a way to determine when it is and when it is not permissible to lie. A closed justificatory loop could leave us blindly adhering to the absolutist principle. If we don't allow at least some cases of deception, we'd be left letting murderers into our homes.

Even many contemporary neo-Kantians grapple with this absolutist interpretation of Kant's ethics. Some claim that rather than rely on our own private mental resources to determine whether a principle is consistent, thereby running the risk that we will confuse our own private wants with the wants of others, we instead need to validate our justifications in the public sphere, which we do by expanding greatly the capacity with which we justify our actions.

TRADING PLACES

It is the sexual fantasy of a great many young men to walk into a room and be set upon by hordes of lusty women, only to be forced, against their wishes, to perform all manner of carnal indignity. Anecdotally, my straight male students seem to heartily endorse this fantasy. "Yep. Sounds great!" Prevalent though this desire of young men may be, it doesn't neatly translate over to the other half of the population. A much smaller number of women share an analogous desire. In my classes—not a representative sample, but still—the number of women excited about this prospect very nearly approaches zero.

Importantly, the men in question are expressing *real* desires, meaning that if such a scenario were actually to come to pass, it is plausible that these men would likely leave the scene feeling as though they had won the lottery. A great many women *fear* precisely this situation. This, again, is a very real fear. The critical mistake occurs in attributing one's own desires to another person who only may or may not share those desires.

I don't want to go attributing desires recklessly to one gender or the other. Cases may span the spectrum. Some young women may secretly harbor this desire, and some young men may fear it. I want only to call attention to the way in which strong motivating desires can be misapplied to other people, because this helps illustrate why we can't go basing our moral rules on the beliefs and desires of others, as well as why it is important to keep our justificatory loop open.

You can change the argument to suit your tastes or to suit the stereotype, but I think the point should be clear: desires aren't transitive. It is a deep mistake to assume that our own desires are shared by others. If you act on your desires and treat another as you would *want* to be treated, you run the risk of trampling their desires. People—you included—desire all manner of crazy things, some of which are personal preferences and some of which are affected by outside circumstance. In turn, this affects the willingness to pay that we discussed earlier. If you are rich and I am poor, we may have different price points for things, entirely contingent on our bank accounts. Whatever price I might accept for my kidney is likely not the same price that another level-headed kidney owner might accept. More importantly, the price of consent is equally nontransitive.

Some people mistake the Kantian Categorical Imperative as yet another version of the Golden Rule. But this is dead wrong. The Golden Rule—"Do unto others as you would have them do unto you"—suggests that you should treat others as you yourself would want to be treated. The Categorical Imperative suggests something quite different. It employs the strong distinction between wanting and willing that we discussed in chapter 5. Whereas the Golden Rule instructs you to treat others as you would *want* to be treated, the Categorical Imperative implores you to act as you could consistently *will* that all would be treated. The imperatives generated by the Golden Rule, in other words, are hypothetical and contingent on our desires, but the imperatives generated by the Categorical Imperative are categorical and noncontingent on our desires.

The hypothetical imperatives of the Golden Rule can easily take the form of operations: "If you want to do X, then you should do Y." The prescription—the *should*—is spelled out by the desire, which is unique to each person. As such, the Golden Rule can be used to justify a great many crimes against others. Just because Jones wants to go swimming doesn't mean that Smith will also want to go swimming. Just because Tina wants it made public that she has gotten engaged doesn't mean that Marcy will want it made public. Just because I would donate my kidney to help an ailing stranger doesn't mean that you would. "That's what *I* would've wanted!" can be used to justify forceful sexual interactions, the unauthorized extraction of kidneys, and the obliteration of sundry natural creatures and entities.

Categorical imperatives aren't open to this sort of distortion. They apply regardless of circumstances, regardless of wants and desires. The Categorical Imperative—capital C, capital I, which is really just a mechanism for spotting categorical imperatives—has its eye on identifying universal rules, or laws, that apply to everyone.

Here's Kant's second attempt at articulating the imperative. You can see how different it is:

The Categorical Imperative, Variation II:

> Act in such a way that you treat humanity, whether in your own person or in the person of any other, never merely as a means to an end, but always at the same time as an end.

The big idea here is that other people ought to be respected as reasonable and rational, not used to help us achieve our own selfish purposes. Again we see that if we do use people to get what we want (and we *merely* use them, neglecting to treat their objections or protests as meaningful), we disregard an important rule of reason: that we cannot make ourselves an exception to the rule. Why? Because that's not what a *rule* is. We would be imposing our wants and desires on others while completely ignoring their wants and desires, assuming that wants and desires are important enough to act upon but cherry-picking which ones to privilege. If we desperately want to have sex with another person, and that person protests or says no or rebuffs us, we are obligated by our own internal logic to respect that objection. If we want our self-lawgiving to be respected, we must issue laws that we ourselves respect. Self-lawgiving, or autonomy, requires this of us. Universalizing the rule that we can willfully use others for sex against their will would mean that others could willfully use us for sex against our will, which in turn would mean that we were simultaneously willfully and not willfully engaged in sex.

Okay, that's fairly technical and confusing. If you don't understand it, don't worry about it. A somewhat related but easier to grasp point is that even well-educated, enlightened adult humans are fairly bad about knowing and/or articulating their wants and needs. Since we're all different in minor ways, the project of cataloguing our wants and needs will remain forever unfinished. Fortunately, most adults maintain relatively sophisticated capacities for communication, so we can articulate and negotiate our wants and needs with one another.

Understanding the wants and needs of others gets all the more complicated when we wander outside the relatively small cluster of flesh-bearing creatures that can grasp and articulate these wants and needs. It's difficult to know what a child wants, for instance, their fleshiness notwithstanding. Children are unformed and often ill-informed wills. If they throw carrots from the table in protest, it is more likely that they simply don't know what they want, or they haven't yet developed a taste for carrots, and so must be guided through the process. Fortunately, kids can improve at articulating their wants and desires, and we can help them with this. When a fight breaks out on the playground, a chorus of parents may chime in with "Use your words."

Animals are even more complicated though they exhibit similar behaviors to little children. They may act out of fear, but we can extrapolate their interests from information we have about them. Plants too respond to environmental circumstances. They may not have wants and desires per se, but they do have needs, and we humans have entire branches of science, like botany and agriculture, devoted to making sense of those needs.

As rational and self-directed agents, we're in a prime position to interpret these reasons, though they become considerably hazier the farther away we get from circumstances with which we are familiar. "Why did the bear choose to attack Samuel Ives when there was an abundance of food that summer?" We really don't know, but we can analyze this behavior and seek an explanation.

Briefly consider what it is to *have* a reason. When the Ives boy was dragged out of his tent, the bear that dragged him presumably had some inkling of what it was doing. In the wake of the attack, reporters and other curious onlookers, as if by instinct, attributed such thoughts to the bear. Perhaps it was a case of mistaken identity, they conjectured; perhaps the boy smelled like a particularly delicious marshmallow sandwich, or perhaps the bear was famished and desired the taste of humans, who had once tormented him. We'll never know. We can only speculate. It was the law of the jungle. These explanatory observations help evaluators better understand what happened. They are accompanied by thoughts of sadness or empathy and a concern about the safety of the wild.

When Randall Lee Smith, in contrast, pulls a gun on innocent campers, our instinctive response is much less forgiving. We immediately leap to thoughts of disapproval, we lay the blame at his feet, we hold him guilty of wrongdoing, and we view him as wicked.

So we can unearth reasons that may be sedimented in our behavior or in our prior commitments. We can even unearth reasons evident from the behavior of others. We can do this with serial killers: "Smith shot the couple because he wanted to rape Susan." And we can do it with animals: "The bear was looking for food." Some are explanatory reasons: why Smith shot the couple or why the bear grabbed the boy. Some are justificatory reasons: "You would love to have a beer, but your grandmother is in the hospital." In most circumstances, a dying grandmother will serve as "reason enough" to pass on a beer. Other cases will be different; the mere statement that you are choosing to go home and snuggle may need further justification: "We've been really busy all week and haven't spent any time together." Depending on the strength of your preferred justification, this either will or will not make the case for you.

The reasons we offer can appear more like an explanation than a reason embedded in the self-direction of the entity. "My orchid loves the light. Put it in the corner." Or even more abstractly, "Lightning wants to get to the ground." Yet these claims about the world, and about how the world works, are not strictly explanatory. We can really learn a fair bit about one another simply by extrapolating from generalizations.

What I've said so far, in essence, is that we need to double down our efforts to justify our actions and that justification can be informed, but is not fulfilled, by explanation. We need to revisit the reasons for which we act and ensure that they can pass justificatory muster. Undue emphasis on the good crowds out these reasons, stops them up, filters them. Surely there is room for philosophical investigation into the nature of the good, but on a day-to-day basis, once we've identified a good, we needn't investigate any further. Identifying and fixating on a good lays the groundwork to make us philosophically and intellectually lazy, to close off our justificatory discourse, and to focus on promoting that good instead of justifying our actions.

PHILOSOPHY, NOW MORE THAN EVER: SECURING OUR ENDS

Drunk driving, like feeding bears, is exceptionally stupid behavior. It is unarguably dangerous. It can be costly. It is risky. It presents not only harms, costs, and risks to oneself, but also to potentially many innocent others. Yet I suspect that any reader who drinks even socially has been stuck at the bar after two or three beers and wondered about getting home. "Would you like another beer?" the waitress asks politely. Yes, of course you'd like another beer. No snuggling tonight!

It is enticing, in these circumstances, to drive. Once or twice, in moments of foggy imprudence, you may have done exactly this. If so, you understand the ease with which one can fall victim to the immediacy of desire.

Thing is, this sort of indiscretion doesn't happen because you disagree with drunk-driving laws. Perhaps, like most, you accept and endorse the view that drunk driving is incredibly disrespectful and stupid behavior. There are many reasons to support drunk-driving laws, after all. Nevertheless, there you may be, in the moment, living your life, stuck at the bar, when you make a decision that under other circumstances you might reject.

It's not the risk or the cost of harm that militates against drunk driving, although the possibility of these outcomes highlights the point. No, the reason that one ought not to drink and drive is that it is important to the practical functioning of our transportation system that we draw some lines at irresponsible behavior. It's the logic of the system, in this case tempered by what reasonable (not rational) people can accept or tolerate. I certainly hope when I get on the road that all drivers are looking out for me, just as I look out for them: again, not strictly because it is that much safer, but because it is deeply disrespectful of others to drive drunk. Improved safety, of course, is the objective of avoiding drunk driving, but the reason you should refrain from drunk driving is that it is outrageously selfish to put your interests in getting home expeditiously over the safety of others.

Many reasonable people are willing to accept other dangers, like a higher speed limit, for instance, even though there may be more deaths as a result of those higher speed limits than from drunk driving. Certainly one possible solution to the drunk-driving problem is to slow the speed limit down to three miles per hour, or to create special drunk-driving lanes with enormous padded walls, where people can drive recklessly without concern that they will harm themselves or others. This would certainly transform the problem into a nonproblem. Most of us don't place priority on drunk driving, per se, so there's no compelling reason to go to these lengths. But we do place priority on drinking and having a good time, so it makes sense for us to create alternatives that make the drinking without the driving a possibility.

A second complication of drinking and driving is that when you're drunk, you can't take much responsibility for whatever wrong you've done—and operating a heavy machine demands that you take responsibility for your actions. If you voluntarily give that up, you need to be sure that you take responsibility for the things you do before you reach that state. There are deep problems with drunk parenting, being drunk on the job, or just plain living your life in a completely drunk way. Almost none of them relate to injury, cost, or risk. You're not a good parent if you're drunkenly parenting; you're not a good worker if you're drunk on the job; and you arguably aren't being a good person if you are drunk all the time and are unable to live up to your obligations. None of which is to say that you ought not to drink, that there is no space for fun; it's just that you cannot grade your students if you are drunk and also be a good grader of students. You can, of course, be a *bad* grader of students, or you can be a *lucky* grader of students—maybe all of your darts hit the dartboard just right—but you cannot be a *good* grader of students.

The important point here is that we can secure our autonomy, secure the rightness of our actions, by creating the conditions in which we continue on the correct justificatory path even when we are unable to do so. If we are drunk or stoned or lobotomized, our autonomy is impaired. We are not, at least as far as our actions are concerned, entirely capable of accounting for what we do. Yet at the same time, particularly if we are only temporarily impaired, we are responsible for what we do because we have likely put ourselves on the path toward impairment. We are still agents.

It would be one thing if recklessness happened only after raucous nights at the bar or after whiskey-filled grading marathons, when we are intoxicated and incapable of making decisions about what is right or good. But they don't. We face a plethora of decisions throughout the course of any given day, and it doesn't take booze to bring us to the point at which we aren't actually making these decisions at all. It turns out that texting while driving can be just as decision-impairing as drinking and driving. Many other distractions present greater or lesser challenges. Chatting on a cell phone, thinking about the events of the day, disciplining a child, planning a surprise, rocking out to Queen—the multitude of distractions and life-complicating factors that each of us must channel and sort and surf over any given moment nudges our attention into a confined and claustrophobic space.

And those are just the distractions. There are many more omnipresent pressures that tug us farther away from justification. Every time I'm at the grocery store and encounter a dozen eggs, I ask myself whether I should buy the organic free-range eggs or the cheaper eggs. Organic eggs are sometimes two to three times as expensive as the factory-farmed eggs, and even for a few dollars it is easy to be enticed away from doing what strikes me as the right thing. So too with higher-octane gasoline, organic coffee, and sustainable hardwoods—almost any responsible thing you care to name. It is cheap and easy to be brutish and immoral toward one another, toward the rest of the world. The price of humanity sometimes costs a bit more.

So this is actually the gambit we face: to ensure that in our off moments we do not mistakenly do the wrong thing. Our freedom and our autonomy are secured by giving ourselves rails to ride on. We need some device, some mechanism, that will keep us on the justificatory path even when we are not paying close attention to the details.

To ensure that I do the right thing and am not tempted to act on bad reasons, I often endorse and support the passage of laws that will keep me from having to justify every purchase. If I know that all eggs are free range, or that animal cruelty laws prevent chicken farmers from unacceptable farming practices, I can more easily make it through my shopping list. I can focus on price, as I'm inclined to do, instead of also asking these difficult moral questions at every decision juncture. I therefore support the passage of laws that keep me from buying a gas-guzzler, for instance, even though I may want one, and that lower the price on an alternative-fuel vehicle so that it is more in tune with the principles I support because they are right. Simply relying on the market to make these decisions is a bit like allowing me to decide on my principles while I am trying to do something else.

As humans, we have reason to be rational. Though it sounds like a tautology, I suspect most readers of this book will agree that it is rational to be rational. True, we are animals, too. We can, at times, deny our human nature. It is sometimes even fun to be animalistic. But as humans we must place rational constraints on our behavior in order to respect rationality, in order to advance our freedom. There are responsibilities we must fulfill, duties that we must uphold, lest we suffer the scrutiny and disapproval of others. Identifying and acting according to rules doesn't place artificial constraints on our behavior, as many assume. The right rules make it possible for us to be rational and reasonable in the first place. It's the job of ethical justification to suss out the right rules.

10

The Good Green Life

In which the author wraps this whole disquisition up into a tidy little paper package.

THE PARABLE OF WICKED AND WILD

Imagine two separate worlds, identical in almost all respects. Each world is filled with people and animals and fungi and trees; decorated with rivers and rain and shadows and clouds; punctuated with buzzing insects and crashing waves and booming thunderclaps. Each is beautiful, to a degree, but each is also somewhat ugly, at times. On balance, each is a difficult place to survive, with scarcity and injury and death around every turn. Save the nastiness, there are moments of pleasure and joy as well. Life expectancy is short—around 40 years—and children sometimes do not make it through their early years. A misery meter—the same sort that tracked misery during the Boxing Day Tsunami and the bombing of Hiroshima—needles in the background, logging pain in quantifiable units, mirroring in reverse a happiness meter that tracks the little joys that each world provides. On balance, misery outweighs happiness in each of these two worlds.

The two worlds differ in one critical way.

In one world, the world of Wicked, people take action primarily for bad reasons: they harbor ill wishes, unjustifiable objectives, and selfish disrespect for others. In the other world, the world of Wild, people take action primarily for good reasons: they harbor well-wishes, concern that their actions be acceptable to a wide range of affected parties, and justifiable respect for others. Whereas Wicked is miserable by virtue of the people who live in it, Wild is miserable by dint of some other feature. It is filled with noxious pathogens, say, or it is prone to a more active and hostile climate.

Imagine that these worlds are completely unconnected to each other—in separate universes, even—though better and worse states of each world occur in parallel. For every evil deed done in Wicked, there is a bad event in Wild. For every bad event in Wild, some evil deed is done in Wicked. If some ill-intentioned brigand pulls a trigger in Wicked, thereby killing an innocent and harmless child, so too does a pathogen jet out of the shadows and kill a similarly innocent and harmless child in Wild.

By assumption, both of these worlds are equally miserable. Wicked is miserable because people make it miserable. They kill and maim and hate and loathe. Wild is miserable just because it is naturally so. Nature is cruel. Try as they might, the people work together but fail at improving their lot. The misery and happiness meters flutter along in perfect synchronicity. When Wicked's meter hits 23 on the dial, so too does Wild's.

The people in Wicked are like our poisoning stranger, like the Appalachian Trail killer, like each of us, some of the time: inconsiderate brutes and animals fumbling through life without a care but for the gratification of our immediate desires. They play games with one another, view life as a game, something to succeed or fail at, something to conquer and beat.

The people in Wild resemble what many of us would like to be, and what we actually are, also, some of the time. They justify their actions, they run their reasons past other affected parties, evaluating the strengths and weaknesses of this or that position. They challenge one another, taking into consideration the effects of their actions on others.

Which is the world worth living in? Where would you rather be? You have only two options.

This world of Wicked: Is this a good world? To my mind, it is a pretty terrible world. It is a deeply unethical world. On this much, I hope we can agree.

This world of Wild: Is this a good world? We've already stipulated that it is not a good world. On balance, misery outweighs happiness. But I think we can also agree that Wild is an ethically neutral world, and because of this, that Wild is a preferable world to Wicked.

I, for one, would certainly rather live in Wild than Wicked. I would rather be a citizen in Wild than Wicked. I would rather face the perils of tar pits and quicksand than suffer the indignities of man against man.

In Wild, life is hell, but people do not wrong one another; they try to do right by one another. It is a terrible place to be, but the citizens struggle mightily to survive against all natural odds. There is no murder, for no one wishes misfortune upon another. There is no petty thievery, for no one pursues the material goods of another.

The problem is that this is just a fable, a parable, an allegory. It's a fiction. The people in Wild have experiences not so distant from our own. The people in Wicked also have experiences not so distant from our own. They sometimes guide their actions, as many of us do, some of the time, out of wickedness.

But there's at least one other world we have yet to consider.

EARTH

Imagine now another world that is just like Wicked, just like Wild, just like both of our previous fantasy worlds in every miserable and happy way but one. Imagine that in this world, the people do not act with malice or ill intent; and they do not act with good reasons. For most of their lives they allow themselves to be dragged along by instinct and desire, by the same constellation of outside forces that drags jungle critters and sea creatures through their Darwinian survival routine.

To abet them, the people of this world have developed an elaborate institutional infrastructure that enables them to quickly and rapidly assess the intensity of their desires and trade between themselves the gratification of their own desires for the gratification of other people's desires. They have devised clever actuarial methods for hoarding and storing their ungratified desires, chits for transferring desires, laws for skimming the unfulfilled desires off the dreams of those who toil and labor, enforcement regimes to ensure the fulfillment of promises built on desires. They have built political systems that reduce reasonable discussion to mere tally taking, simple tabulation, so that engagement with the civil sphere amounts to little more than choosing which team to support, which jersey to buy. They have constructed enormous media superhighways that flatten reasonable discourse into exchangeable factoids and titillating trivia, transforming the grand human project of enlightenment into the complicated but relatively uninteresting project of data accumulation and knowledge logging.

The people of this world have few identifiable reasons for doing what they do. Their reasons, when identifiable, are plainly roped back into their immediate needs, to the gratification of their most animalistic desires. At times their behaviors appear to be good actions, bringing about good ends. Indeed, the wealthiest among them smile and play with relative disregard for the hungry and destitute who lurk in the shadows. Other times, try as they might, the world lashes back at them, destroying their homes and robbing them of their health. Their actuarial infrastructure and technological expertise permits the pooling of resources, offers an imperfect means of combating the rapacious and unforgiving forces of Mother Nature. They can cash in and build anew, softening the blow.

In this world, the things they do are merely good or bad—for them, for their victims, for nature. It is, naturally, not a zero-sum relation. Some relations are mutually beneficial and some are mutually detrimental. The symbiosis permits mutual harmony of desires and self-preservation. Value is created. Value is destroyed. The wheels of industry spin onward.

If an alien sociologist were to appear in this world, documenting the interactions of its inhabitants, he might tell fantastic stories about human behavior: "These humans, they buy and sell objects to decorate their nests. They set laws to coordinate action. They toil day in and day out, working to accumulate currency that they can exchange for more objects." In so doing, this alien sociologist could describe with great accuracy the sordid state of affairs on this dying planet.

How so?

Because the environmental predicament in its current instantiation is perfectly understandable from the standpoint of description—bereft, as it were, of justification.

The core environmental problem isn't that we don't value nature. Sometimes we do and sometimes we don't, all for varying better and worse reasons. The core environmental problem is that we often don't think about what we do, and we've built an enormous machinery of reinforcement mechanisms that make it easy and convenient to avoid thinking about what we do, that in fact *shield* us from thinking about what to do.

When I say that we don't think about what we do, I mean so in a very specific sense: we often act without justification. We don't do what is incumbent upon us as human beings. Almost the entirety of our economic life, of our daily political life, of our routinized nose-to-the-grindstone life, is oriented toward quashing our justifications before they get off the ground.

Questions about family, health, safety, security, freedoms, and rights pervade every environmental issue. At present we answer these questions by permitting outside forces to determine the playing field on which options are presented to us. Our front-line environmental soldiers treat every environmental issue like a battle to be won. Even the environmental diplomats, try as they might to persuade industrialists and marauders to do otherwise, revert to rhetoric and strategy in the service of achieving some end.

Environmentalists for decades have been trying to shift the focus from Wicked to Wild, to suggest that we must force the discussion onto the beauty and/or the value of nature—to recognize once and for all that nature is not merely a beautiful backdrop, a beast to loathe, a land to conquer, or a pool of resources to utilize. They have sought to show that there is value there; that we must be cured of our misdeeds and misapprehensions; that we must be shown the light and driven from the darkness. Perhaps this is the natural response to human activity that systematically runs roughshod over nature, but it's not necessarily the correct response, and it sure as heck isn't the most practical response.

What many have failed to do is to make the case for justification. They have failed to acknowledge that just as we are not living in Wild, so too are we not really living in Wicked. They have offered justifications, yes, but for those who believe that they can follow the laws of nature, or that nature's value consists in its use to us, or even that nature is valuable in itself—those who believe, in other words, that all we need to do is focus on better and worse states of affairs—the justificatory loop is closed. We need only identify the better state of affairs and then our marching orders are clear, provided we agree that good states of affairs offer marching orders.

I think it is clear that good states of affairs do not, in themselves, offer marching orders. To get to marching orders, we also need to investigate seriously the notion of justification and responsibility, and we need to acknowledge, above all, that our marching orders come not from without but from within—which is all to say that they're not marching orders at all, but instead self-legislated expressions of our freedom and rationality.

FOURIER'S BLANKET

In the exact same year that our dear friend the Very Reverend William Buckland lots-of-important-initials-signaling-erudition-and-prestige-after-his-name christened the erstwhile owner of the Scrotum Humanum the megalosaurus, thereby writing a new page in the paleontological encyclopedia and causing his entourage of proponents and detractors to enter a tizzy of replies and rejoinders about great floods and major upheavals and uniformitarianism and catastrophism—in that year, in 1824—another extremely impressive scientist and meddler, Jean-Baptiste Joseph Fourier, mathematician and physicist, progenitor of the Fourier series and several other important mathematical proofs—folks really dabbled back then!—discovered and published the first paper on a phenomenon has been used the world over to grow tomatoes and flowers and courgettes and rosemary, but until that paper was written and published had otherwise been unassociated with the geophysical dynamics of the earth.[1] The greenhouse effect, as he called it then, and as it is still called today, describes the state of affairs that ensues when you spit a bunch of crap into the atmosphere and trap sunlight between the soil and the sky. It is no hypothesis. It's a real, demonstrable effect.

Fourier's genius, sadly, would ultimately undermine itself severely, for it was only a few years after discovering the greenhouse warming effect that Fourier also concluded that warming the human body, and keeping it warm, was conducive to good health. To keep himself warm, he routinely swaddled himself in blankets. With irony that perhaps only Jonathan Franzen could persuasively relate, one fateful night in 1830 the father of the greenhouse effect wandered through his Paris flat, tripped on the corner of his health-promoting blanket, and fell down the stairs to his death.[2]

It is sometimes said that cynics know the price of everything and the value of nothing. I have argued above that our reasoning can be, but ought not to be, collapsed into a conception of the good. When we transform our reasons into a monolithic idea—like the value of nature, the worth of a life, the price of a product, the pursuit of happiness—we winnow our reasons down into that single, univocal measure. We reduce the complex relationship that we have to the world and that the world has to us. This is easy to see, and equally easy to miss, if we abstract away from the complexity of our decisions in order to model them as simplified decision junctures, which I discussed in chapters 4 through 6. It is more helpful, as I suggested in chapters 7 through 9, to turn our attention to a different means of justification.

I have also suggested that the unreflective pursuit of the good, an orientation toward success, undercuts the very process by which the good holds its place *as* good. If Jones adopts the attitude that the world is something to be conquered, life a game to be won, then he effectively interprets the rules as *mere* constraints on his actions. In games where the rules are not explicitly spelled out—which is to say, in nature, red in tooth and claw—all actions are fair. This fact alone, to my mind, serves as an indictment of the orientation toward success. All actions are not fair. We are bound by our rational nature as human beings to be moral, to respect others and the world we live in.

To uncover the rules that we are bound by our rationality to live by, we must adopt a different attitude and orient ourselves toward justification. This more philosophical attitude should encourage deliberation and reflection by all affected parties about what the rules are. We ought to be doing this for every decision we make.

The peculiar literary decision I referred to in chapter 1 is therefore not that I began a book on environmental responsibility by invoking the extinction of the dinosaurs, but that I waited until the second-to-last chapter to start steering you in a more positive philosophical direction. This is by design, of course. I did this because our environmental challenges, whatever they are, are not what they may first appear to be.

They are not, in other words, a problem of having our values all twisted up and backward (though in some cases fallaciously held values may contribute to bad actions) but of a refusal to adequately and responsibly interrogate our reasons—a failure to understand that this is the core responsibility we all share, that all of our duties follow from the requirement that we act for good reason. And it is this phenomenon, that we largely aren't acting for ethically robust reasons in the first place, that is at the root of a great many pending environmental issues.

Many in the environmental community, philosophers and otherwise, routinely fail to understand this. They instead persist first with the second step, arguing for a particular conception of the good or a specific value in nature. This is a moderately fruitless exercise.

You simply can't argue for the value of nature and expect that this value will bear fruit if the people with whom you're arguing don't acknowledge that their core obligation is to listen closely to, and participate directly in, the justification of their actions. The problem is not, in other words, that some people have the wrong values, or that others have the *right* values, but that many people think they don't have to do the hard work of justifying their actions, values, and principles at all. They think that all of our normative work is done by science, or economics, or technology, or law, which is just false.

Our normative work cannot be done by relying on the standard goods—happiness, health, money, nature, freedom, Elvis—nor by adopting the primitive and animalistic orientation toward success. To assume that it can, as many environmentalists often do, is to commit Fourier's mistake, to trip on our own blanket. Our normative work can be done only by an earnest and honest attempt to make clear conceptual progress on the rules that enable us to escape the bonds of nature. *Denying* our human caught-upness in the moral project, as many environmentalists have been inclined to do, is to stack the deck in exactly the wrong direction.

The point of ethics, so far as I can tell, is not simply to identify what we ought to do but instead to give us insight into *why* we ought to do what we ought to do. It is important, no doubt, to identify what is good and bad—and those who argue for the value of nature or the scientific state of affairs make great progress in this respect—but our answers must always be accompanied by an attendant justification as to why we ought to do the right thing.

It is no easier to know and understand what the right thing is than it is to *do* the right thing. Every time we are at the store we must make a decision about what to do. Should we buy the free-range eggs or the less expensive eggs? As I mentioned in the previous chapter, I find it extraordinarily difficult to spend an extra dollar or more on a product when there is a nearly equivalent but cheaper product sitting on the shelf beside it. Price is not a trivial consideration. It is a real, serious consideration, and even the most well-meaning and concerned individual will fall victim to it. Everyone wants to be a smart shopper. This compulsion isn't limited to a problem of action, in other words. It creeps into our conceptions of what's right.

The philosopher R. M. Hare once asked whether slavery could be justified.[3] Most alive today believe, of course, that it cannot be justified, and Hare thought this too, but he offered a "'what if'" scenario: What if, in the imaginary lands of Juba and Camaica, slaves were made better off because of their slavery? Would this be enough to justify slavery? As one of the preeminent utilitarian do-gooders, he was committed to the idea that slavery *would* be justified in those cases, but instead he decried those cases as mere fantasies. Slavery of that sort would be nothing like the slavery we encounter in the real world. The facts matter, he insisted. He got around the theoretical implications of his view and insisted instead that his moral position was the only moral position subservient to facts;

that clearly, obviously, only in the fantastical worlds of Juba and Camaica could such an upside-down slavery produce more good than harm.

The problem with his protests should perhaps here make itself clear: Then as now, many slave owners have argued exactly this, that they were improving the lives of their slaves. Slave owners weren't always or only acting in their own best interest. They also believed that they could justify their ownership of other people on grounds that they were making their slaves's lives better.[4]

I think this is completely wrong. Even if our products would be cheaper and our slaves would be well cared for, we must ask the important question about the permissibility of slavery before we ever open the door to the question of whether the price or the value or the overall good in the world matters. More than this, however, because a huge percentage of the population can be incredibly wrong about what counts as justified, we must consider this project of justification always unfinished, forever fallible, and leave the door open to correction. For it could be the case that at some point in the distant future, a careful critic saunters up to a podium or stands on a soapbox and tells us that we are wrong to consider slavery permissible, no matter if the slaves are made better off.

So the standard objection that we must decide between trees or jobs isn't just misguided, it's totally absurd. It's not that it won't be difficult in some cases to constrain our behavior. It's not that it will be inexpensive or prohibitively expensive. It's not even that it will create or cost us jobs. It either will or will not create or cost us jobs. It's that whether it will create or cost us jobs, whether it will preserve nature, cannot be the first question we ask. I am talking here about the framework in which we reason about our decisions and not so much about the conclusions of the decisions themselves.

Pro forma meetings of government representatives from the US Forest Service and the US Bureau of Reclamation, political grandstanding on the part of self-interested politicians, greenwashing from corporations—these are all smoke-and-mirrors attempts to obfuscate a true project of justification and endorsement. The more we allow these kinds of activities to permeate our discourse about the environment, the more likely the whole thing is to derail.

We humans are not wicked or bad or evil, but we are, for the most part, irresponsible. We are irresponsible in allowing the world to swarm around us, denying our humanity, denying that we must answer to reason. We are irresponsible in pretending that our predicament can be resolved if we return to our animal nature—lemmings or frogs or whatever—and in refusing to engage our reflective capacity, in refusing to deploy the vast cognitive resources that should enable us to work through our problems.

What we need is a ground-level project of discussion and deliberation on a scale that Tocqueville could never have anticipated, that the founders of the Enlightenment envisioned when they threw open the curtains of the Dark Ages. We need not just a project about what will make my life, or your life, or some other stakeholder's life, better or worse. We need everyone to study ethics and philosophy and science.

We already have the tools for this justificatory discussion. We now have vast resources through which to make this discussion a reality. We can communicate at the speed of light. We can investigate through online sources and person-to-person networking. We have departments of universities that are each dedicated to some aspect of this discussion. To some extent, this justificatory discussion has been occurring since the inception of the environmental movement.

The one link that's missing in this grand collaborative project of deliberation and reflection is the general reader, the layperson, the plebe, the foot soldier; the butcher, the baker, and the candlestick maker.

PUBLIC HEALTH MENACE

A surprisingly prevalent criticism of environmentalism begins with the suggestion that Rachel Carson, or some such paradigmatic icon—Paul Ehrlich, Al Gore, Dave Foreman, John Muir, Aldo Leopold, the list goes on—is responsible for more death and more misery than the environmental problem that he or she so heavily criticized. This argument is fallacious in many respects, but not least in what it implies we are to assume about environmentalism as a general platform. It implies that environmentalism runs counter to concerns about humanity, counter to concerns about jobs and growth and money and well-being. Consider Carson's legacy.

Critics encourage us to look at cases of malaria in Sri Lanka. Before the publication of *Silent Spring*, they say, public health professionals were working feverishly to prevent the spread of malaria through the use of DDT. Many indicators suggest that aggressive DDT efforts were working. Immediately after the publication of the book, they suggest, the use of DDT dropped substantially, in part because hypervigilant environmentalists, spurred by Carson's romantic pining for songbirds, were selfishly on the lookout for ridiculous connections between environmental hazards and their localized welfare.[5]

In response, vanguard environmentalists leap to Carson's defense, spouting numbers and data that call the arguments of the anti-Carsonites into doubt. They point out that Carson herself didn't entirely oppose the use of pesticides but rather felt that they should be used sparingly and after the appropriate research had been conducted. Without question, this is a legitimate response, perhaps even aimed at the sort of justification that I'm advocating here, but without interlocutors who also feel the pull to act for justified reasons, these defenses of Carson fall on deaf ears. Those who defend her are bickering over the facts when what they really must do is suss out the norms. The first, and in many ways prior, challenge for environmentalists, as I've been saying, isn't to defend the beauty or wonder of nature, it's to get nonenvironmentalists to see that they too must act for good reasons—that is, reasons that pass justificatory muster.

Whether Carson's book is what caused the use of DDT to stop is an open question. Some immediately leap to that conclusion, where others speculate that the diminished number of malaria cases led policy makers to pull the plug on spraying. Since I'm not an epidemiologist, I'm not one to judge, but my suspicion is that it may have been a combination of the two causes. Seems perfectly reasonable. After all: the number of malaria cases had dropped to the point that they were negligible, so why continue using a pesticide that apparently has some ill effects?

Just as vaccination programs have caused a debate about the responsibilities accruing from them, the use of DDT inspired a debate about what is right, not only about what is good. It is important that part of this debate involved a discussion of the good, about whether the human health effects from DDT are of concern when compared with the human health costs of prolific mosquito populations. Yet the nature of the good can be appropriately understood only from the perspective of this open discussion, not from the perspective of what is better or what is worse.

I suspect that there are various ways to read Carson's work and understand what she was up to. If viewed as oriented toward success—that is, by those who are skeptical about the motivating reasons of environmentalists—Carson's work likely appears directed primarily at moving people to change their behavior. It is easy to attribute all sorts of nefarious intentions to Carson. She was pursuing her own selfish end or chasing some pesticide-free utopia where antiwar peaceniks and tittering songbirds can live in fantastic harmony.

Adopting the lens of the success orientation not only opens the door to these ludicrous interpretations of legitimate and earnest work, but it also works in the service of furthering rhetoric. Because the interpretation is so persuasive (ah, yes, I see now what she was up to; she was chasing an ideal), it then becomes easy to debunk.

If, however, one attributes to Carson an orientation toward justification—an attempt to articulate, communicate, and relate a problem so it can be addressed collectively and cooperatively—then this, I think, begins to seem like a much more worthwhile project. *Silent Spring* initiates a discussion about whether the use of pesticides is justifiable, in which what is best is quantifiable not in terms of the number of songbirds or the rate of malaria mortality but in terms of what we can justify. Carson was making an argument. In many instances it is hard to justify the rampant application of DDT, but in other cases it is easier to justify—we are talking about 250 million annual malaria infections, resulting in 1 million deaths, after all.[6]

Because the emphasis on end states of the world is on the getting there and not on why we should get there, all justifications that appeal to end states have the appearance of, and to some extent are driven by, the orientation toward success. What we need is a clear framework from which to determine whether an action is justified. This framework can be found, I think, if we understand justification as a process through which a decision must go. If the decision leading up to the action has gone through this justificatory process, then the state of the world that results from the process of justification can also be called justified. Even more than this, hopefully this justified state of the world will be the good state of the world that so many people desire.

If this is not the result, then new possibilities open up for us as decision makers. Perhaps we must revise our initial assessment of what state of the world we are striving to attain. Perhaps we should return to the affected parties and reevaluate our facts. Whatever the case, justification is an unfinished project, always a work in progress, and with every action we take, with every movement of our arms, with every check we sign, with every policy we enact, we are called upon to ensure that our actions are justified.

HAVING A REASON

Some people will find what I've said here to be tremendously unsatisfying. What I've said seems to have no teeth, to offer no guidance, to be too open-ended, to be empty and devoid of content. But I think they'd come to a different conclusion if they thought a little differently about what sort of guidance they're currently getting.

If I say, "Do whatever confers the most benefits in general; maximize welfare," I am offering a guiding principle to live and set policy by. I haven't laid out a menu of items that maximize welfare, nor have I told you what to buy or what to avoid. We still need to have that discussion. So too if I tell you to maximize profits or GDP.

If I say, "Do whatever maximizes benefits to the minimally well-off," as was recommended by John Rawls in his monumentally influential *A Theory of Justice*, I am offering a "maximin" principle. I am again offering a policy-guiding principle. I chart no course, sketch no map, write no text that clearly spells out what to do.

If I say, as Michael Pollan says, "Eat food. Not too much. Mostly plants." I have still not given much of a prescription. Likewise, if I parrot Dave Foreman and say, "No compromise in defense of Mother Earth" or if I say, echoing the great disobedient hermit, "Break the law," I have not specified when not to compromise or when to break the law.

If I say, as I do here, "Do what you can justify. Reflect on your actions. Ensure that they can and could apply universally, without contradiction, by running your reasons past friends and neighbors and your toughest critics," I am also offering a policy-guiding principle. It's a principle of justification, which I think should be the core tenet of environmentalism.

Every time we act, we act for one or another reason. We need to make sure that the reasons for which we act are good reasons, valid reasons, *justified* reasons—reasons that have gone through a justificatory process that confirms that those reasons could withstand the scrutiny of outside parties. The current state of affairs is exactly upside down. We need little or no justification to take action. We simply act on our desires, on our preferences, based on what we want, however those preferences were formed, however bottomless our bank accounts are. We purchase a new television because this is what we want, because it is a good deal, often without a thought about the carbon footprint of that television: of the heavy energy load that will be required to run it, of the minerals that have been mined to build it, of the human labor and the cost of the labor that has gone into piecing it together, of the site where it will eventually degrade and decay once the next greatest gadget comes along to take its place.

Does this justificatory position have content? Well, yes and no.

The content comes from the context; the facts and norms particular to each action we take. Where we are. Where we've been. What we're doing. What we expect.

Surely, when it comes to justifying our actions, we need to appeal to consequences, to good and bad states of the world. But these states of the world are not doing the justificatory work. Perhaps you think that a clear-cut forest is bad and that clearcutting a forest is therefore wrong, or perhaps you think that global warming is bad and that polluting the atmosphere is therefore wrong. No question, these things are extremely problematic—maybe quite bad. It is important for us to account for goods and bads in the world. But in my view, the error is that in many cases we're doing these things for the wrong reasons, making very little attempt to ensure that the reasons for which we're acting could be defended to others. We clear-cut forests to make paper bags and cardboard trays that we toss in the garbage after one use; we pollute the atmosphere to make fashionable widgets that pop in and out of style or that we use for only a short time.

What makes these actions wrong is that much of the time these mindless, thoughtless activities—consuming just to consume, to satisfy a craving, to fulfill a perceived need, to conduct "retail therapy"—essentially go unchecked and unscrutinized. Cutting trees to build shelter, to make a fire to keep ourselves warm, to grow food is much more easily justified (meaning defended and/or defensible to the wide public community of which we are a part) than cutting trees to create cardboard boxes so that we can ship foods from one side of the country to the other.

Instead, we must agree among ourselves—as human beings, citizens of the world—which policies and actions are justifiable. As the sole bearers of responsibility on this planet, we have to ask ourselves: "Is what we have done justified?" I'm offering a method of determining whether your action is justified. Round up your reasons and take them on the road. Take them to your friends, who will likely agree with most of what you say. Take them to your neighbors, who only may or may not agree with what you say. Take them to the person on the other side of town, on the other side of the state, on the other side of the world. Ask them what they think. Take them farther, looking backward, as if from a time machine far in the future, to see whether what you're doing is something that future people could accept. If so, why? Be careful, as you do this, to distinguish motivational reasons from justificatory reasons. Ask about the justificatory reasons; the motivational reasons won't be helpful here.

When the Mary Annings and Gideon Mantells of the future unearth our dusty old bones—the rusting carcasses of our automobiles, our tractors, our washing machines, the empty caverns of our prisons and our bank vaults—will they too feel that they have stumbled on the lairs and nests of dragons? What story will they piece together long after we have come and gone? Will it be a simple explanatory tale, told to every four-year-old? Or will it also involve some justificatory dimension, something that lays blame at the feet of the people who lived and walked the earth thousands of years prior, during the 20th and the 21st centuries?

There can be little question that time will outrun us. If not in the next 100 years due to climate change, then in the next 1,000 or 10,000 years due to some unimaginably horrific and awful event. The question is this: Is ruin the destination toward which all men rush, as Garrett Hardin once succinctly charged? Will the cause of our extinction come about by our own hand? Or will our extinction be something only and merely devastating, something only and merely catastrophic?

If residents of our future planet could hop into a time machine and return to our time, how would they understand our predicament? Would they say that what we are doing is justified? Would they forgive us? Will they look to our libraries, our philosophical works, our ethical battles, our intelligent commentaries? Will they see us as driven in large or in whole part by greed, by self-interest, by self-preservation, just like the great beasts that were driven to extinction by a thundering asteroid? Or will they see us as reasonable creatures, struggling mightily to make the universe a wee bit less miserable only to be mercilessly wiped from the stars by the caprice of nature?

Though this view about justification may seem implausible, idealistic, and new, the idea is in fact very old. We have been justifying our decisions for an extremely long time. Fortunately for us, we already do have the resources to address these problems, just as we always have. We have always solved our problems by appealing to factors other than the good, by appealing to rights and responsibilities. We have, in the past, won our ethical battles by moving legislation along according to justifications. The women's suffrage movement, the labor rights movement, the civil rights movement, the gay rights movement, and many others have been advanced and have succeeded not by demonstrating the greatness of women or the value of gays, for instance, but by getting people engaged in the justificatory struggle.

None of these accomplishments were achieved by pointing out the economics of abolishing racism or sexism. There have surely been real political battles, and even some actual steel-and-flesh battles. The Civil War was a bloody mess; the Iraq War continues to be a bloody mess. But there have also been velvet revolutions, brought about through the slow shifting of public sentiment, through the articulation of reasons, both justificatory and descriptive, that brought people to see the error in their thinking.

Not so long ago, during our most recent period of sustained economic growth, from 1994 to 2007, we would read on a weekly basis about new and ridiculously costly desserts and foods: the $25,000 hot chocolate, topped with gold frosting and fresh truffles; the most expensive pie ($14,260), the most expensive cocktail ($1,431), and the most expensive frittata ($1,000).[7] It was ridiculous and frivolous. All but the most unreasonable among us saw it that way.

My grandmother, a survivor of the Depression, instinctively chastised such lavishness by reverting to "need" talk. Her view of need, of course, was relatively thin. All we really *need* is a small 400-square-foot apartment, some blankets, and a hefty plate of liver and potatoes. Characteristically she'd complain that we grandkids didn't *need* so many toys, that we didn't need so many things.

You might be inclined to agree with her and to think, as many people do, either that we can specify the things that people need by establishing a baseline—people of weight W and activity level A need C number of calories a day—or that need is entirely contextual. I think it's somewhere in between. It would be easy to break this down into a question of need—we do not *need* offensively expensive cups of hot chocolate—but the idea of need is notoriously difficult to specify. It draws a discrete line without clearly specifying when and how it is permissible to cross that line.

Think instead in terms of ease of justification. A $25,000 cup of hot chocolate is quite a bit harder to justify, it seems to me, than a $2 cup of hot chocolate. Both are luxuries. Neither is needed. But one is much harder to justify than the other. Need is important here, for sure, but it is not the only consideration. That $25,000 could go a very long way in addressing much more pressing concerns, and to spend it on a single hot chocolate is an unreasonable indulgence—unless, that is, the proceeds from the $25,000 cup of hot chocolate will be used to give 25,000 children 25,000 bowls of rice; in which case, drink your hot chocolate, brother. Buy two. Buy five. Buy hundreds, for all I care.

Or think of it this way. Driving a gas-guzzling SUV is often viewed by environmentalists as a sign of ecological arrogance. This seems reasonable, until you consider that driving a gas-guzzling SUV actually burns less fuel than driving a cement truck or a tractor trailer. So why is one viewed with disdain while the other two are not? It may seem strange unless you consider the concept of need. Drivers of SUVs generally don't need to drive them, whereas drivers of cement trucks and tractor trailers generally *do* need to drive them. But that's not as helpful as one might think, since in all likelihood the need of driving cement trucks and tractor trailers is still tied tightly to the luxury needs of those who want cement and whatever it is that tractor trailers are delivering. Need alone is a bottomless justificatory pit.

What do we need to do? We need to pause at every decision juncture and think hard about our actions. We need to investigate the implications of our decisions, ask what will happen if we follow through on them. We need to ask what sorts of resources have gone into producing whatever it is that we're purchasing. When we walk into a store and decide to make a purchase, whether we actually need something or we just want an object to have and to hold and to cherish forever and ever. We need to justify what we're doing.

So costs and benefits and even needs really do play a fundamental role in this questioning process. We do need to think hard about the costs and benefits of our actions, but these alone cannot be our guiding principles. This is what I tried to show in the previous chapter.

So too for scientific inquiry. We need the sciences to help us better understand what's going on. This was the case in chapter 1 when I discussed the climate sciences. I said there that science cannot guide us or tell us what to do. What science *can* do, however, is offer us information and aid in making our decisions. So long as we bear in mind the distinct normative considerations that must also play a role in our decisions, science has a place.

Some supermarkets have recently taken to asking shoppers at the checkout counter whether they would like to round the price of their bill up to the nearest dollar and donate the remaining pennies to some cause. This is a clever strategy for encouraging shoppers to think about external considerations like human rights, environmental justice, and even the needs of the less fortunate; it puts the shopper in a position that he should always be in. We should be asking ourselves at every juncture whether the actions that we're taking are justified. The only difference is that we ought not to rely on the cashier to press us in the direction of greater reflection. We ought to be motivated to take that burden on ourselves. We should be driven to do this. We should make consideration of others and reflection on our purchases our daily practice, a normal part of our lives.

Really, I think insisting upon the reasonableness (tested by the real but also hypothetical justification procedure I've outlined above) of our decisions is the only way we'll ever get out of the fix that we're in. It's the only way we'll be able to overcome the catastrophes that are projected to fall upon us. Even if we don't actually succeed in changing the ultimate outcome, we owe it to ourselves and to one another to strive to justify our actions, because we are the only entities that can do this sort of thing.

The tendency in the environmental discussion is to focus on the science or the economics, as if either one will resolve our conundrum. But science and economics are very narrow justificatory regimes, and it is clear that they alone cannot suffice to drag us out of the weeds. Not only must we gain insight into the various methods of justifying our actions, of making clear and informed decisions about the right course of action, we also have to accept that this is our responsibility and that this responsibility stems from our peculiar situation as rational and reasonable animals. That's mostly the case that I've made in this book.

Some people will argue, earnestly, that it is our prerogative to answer our *why* questions however we want, essentially offering a shrug and a "just because." This is the logic of antireason, the logic that proposes that we can answer our *why* questions by turning our backs on the questions altogether. They'll say that it has something to do with freedom, that it is freedom-hating to answer otherwise.

But I answer that this is not freedom-hating in the least. Being subject to the whims of nature is itself a form of tyranny; it is to relinquish our freedom to the determinism of nature. Not needing to give a reason for our actions, not needing to offer a justification for the things we do, is in no way emblematic of freedom. It is, in fact, the opposite of freedom. It is being subject to the laws of the earth, like the animals that crawl and scurry around us.

The green stance is the human stance, which demands that we take responsibility for our actions, that we justify our actions, that we consider the needs and interests of others and seek proactively to accommodate these considerations into our policies.

THE VIRIDIAN COMMONWEALTH

A fair portion of this book has emphasized the nastiness of nature, the cruelty that Mother Nature can inflict when we turn our backs. But it has also gone into some depth on the wickedness of humanity. Flip these observations on their heads and we find, at last, the good in nature and the right in humanity. It has emphasized in part how many understand this as a battle between humanity and nature.

We have discussed a variety of horrible dimensions of nature: geological upheavals, plant and mushroom poisonings, animal attacks, epidemic pathogens, self-destructive cancers, weather events, and even psychological misfortune. Equally, we have also seen the range of horrible aspects of humanity: murder, rape, theft, cheating, mass destruction. The important point of this book is that you needn't love nature to be an environmentalist, just as you needn't love other human beings to be a humanitarian. Doing good deeds, being a good person, doesn't emerge from a love for others. It emerges from living rightly, from making sure that you have done your homework. It emerges from being a good citizen.

Contemporary environmentalism has tended toward a velvet revolution, toward packing daisies into the bayonets of soldiers. "Love thy mother" has been the reverse-battle cry, and it has hit home only with those who are soft at heart. But this misses a critical step.

Doing the right thing, acting for the right reasons, emerges from the demand that we be reasonable, that we act reasonably, that we behave as we human beings are capable of behaving. There is a great deal of debate about what morality demands, but if there's one thing that seems clear, it's that morality demands that we be as human as possible, that we accentuate our personhood and recognize ourselves as expressly distinct from nature, as carrying a burden the size of our planet. None of the horror stories I've covered, of course, are intended to scare you away from nature. By all means, go out and enjoy what the world has to offer. There is some fantastic stuff to see and do. I'm planning a long camping trip, including an edible mushroom foray, as soon as this book manuscript lands in the publisher's hands. But it should certainly get you to think about how you can stop being too much of an animal.

Being a good environmentalist does not necessarily mean that you have to become the hippie tree hugger of political caricature. What it means is that you should support pro-environment public policies and that you have an obligation to be involved in the writing and passing of these policies. It expressly does not mean that you have to sacrifice all your desires for the sake of the earth.

We each must understand ourselves as operating according to responsibilities that could and must be justifiable to everyone, for reasons that nobody could reasonably reject, as opposed to seeking out the best and greenest new world. Humans bear a unique moral burden as the only creatures that can act according to principles and justify their reasons. Environmentalists should welcome this moral burden, not distance themselves from it by telling everyone to get back to their natural roots. The best way to determine whether our principles and responsibilities are right is by subjecting our reasons to the widest possible scrutiny, ensuring that they are justified. We can only ensure this, our principles can only be called justified if, and only if, they have been subjected to this wide scrutiny. The insight and input of others is the best check against delusions that stand poised to undermine the green movement.

The most sensible and rights-respecting way of being green, I believe, is the way that acknowledges human beings as special, that does not besmirch humanity by emphasizing its naturalness, that does not strive to reshape humanity in nature's mold, but rather that acknowledges that humans stand in a particularly tricky relationship with nature. Namely: as humans, we can and should seek justification for our actions. Any environmental position that advocates strong-arm policies based on projections about what will be best or on speculation about the good or on an ill-informed deification of nature is misguided from the outset. Instead, the central moral imperative that we all must follow is the imperative to *act with a reason*, to act with a strong and justifiable reason. To do this sufficiently, we must subject our reasons to the scrutiny of others.

Our best hope is to strive for a Viridian Commonwealth, a sort of environmentally informed hypothetical community. Since we are human beings, we must understand our human responsibility if we hope to be true green citizens. We have to begin thinking of the distinction between nature and humanity in the moral terms of the good and the right. This idea is fundamental not just to morality but also to our freedom and autonomy. Since we generally assume that we are free to act as we wish and are not tossed around by the forces of nature, it is essential for us to maintain a distance from nature as well as to recognize that we are caught up in it.

More important, we must acknowledge one critical limit on our freedom: we are not free to do as nature does. If there is a single prohibition that is definitive of our choices, it is that we are free to act ethically, to make ethical decisions, *because we are human*; we are free because we do more than nature does. Any expression of this freedom that has us reverting to animalistic behaviors—acting like brutes—is distinctly not the ethical solution to our environmental problems.

Will the world I imagine be beautiful and shiny and a perfect place to live? Probably not. It won't be anything like that. There will still be misery, heartache, lost loves, tears, death, anguish, and all the sordid horribleness that has for millennia accompanied the human experience. The world I imagine will continue to be difficult. It won't be much easier than it is now. With luck, however, it will resemble Wild more than Wicked.

The upshot of all of this is that you should live life like you mean it. You have an obligation to be a grown-up, to live up to your humanity, to act according to strong, justified reasons. At the same time, you must recognize how easy it is to be pulled away from these reasons by your appetites, by your wants and desires. And sometimes ... this is okay. You may want some new technological gizmo—an iPad, a new car, a fancy grill, a brand new house. Everybody has these desires. Everybody understands them. Within reason, everyone is entitled to act on these desires. You just need to ask yourself whether you can justify getting that new gizmo before you run to the store and purchase it.

You should strive to be a green citizen, someone who takes into consideration all dimensions of a given problem and thoughtfully acts according to what is right. You should do this not because you love nature or think it valuable, or precious, like a vase. You should do this because you are a reasonable, responsible person. Life is not a game. People are not pawns. Nature is not an obstacle course.

If we take ourselves seriously, if we understand ourselves to be reasonable people, then weighing, endorsing, promoting, and responding to reasons is our responsibility.

Postscript: The Right Thing for the Right Reason

The summer of 1967, the Summer of Love, was a time of increasing domestic unrest. The civil rights movement was well underway, Martin Luther King Jr. was in his final year of inspiring young upstarts to object peacefully to discrimination, the mantra that one should "make love, not war," was emblazoned on hand-held posters and tattered bedsheets, and millions of mattresses, shorn of these selfsame sheets, were laying witness to the sexual revolution. We now know that the young man who was snapped by photographer Bernie Boston placing carnations in the barrels of automatic weapons was George Harris, a struggling actor heading off to San Francisco. Harris would later declare his homosexuality, assume the name Hibiscus, and help start a flamboyant drag troupe called the Cockettes. In the early 1990s, he would die of AIDS, a disease he contracted, arguably, thanks to the free love that had become a defining feature of his lifestyle.

It is one of the great misconceptions about environmentalism that one must love the earth—that one must hug trees, embrace animals, grow one's hair long, eat rabbit food, and roll in the dirt—to truly be green. There's no clear genealogy of this idea, but a compelling political story is that it emerged from the confluence of political events in the 1960s. There is little question that the modern environmental movement has its roots in the era of the hippie. Perhaps it is that notion, then, that steers the idea that one must love the earth. This, again, is an explanatory tale; it tells us only half the story.

A somewhat different story makes itself plain if one views the emergent environmental sentiment from the standpoint not of politics but of ethics. That is, the confluence of forces was a mere historical contingency. What really happened is what often happens in cases of profound moral importance: people caught a whiff of the conceptual failings of the then-current mode of thinking and made the rash inferential leap of confusing disregard for nature with disvalue of nature, then immediately set to the task of prettying-up a forgotten heirloom by filling it with roses.

So the answer to our prefatory puzzle at last emerges: the reason that you do not have to love nature to be green isn't that nature is necessarily bad, but that love and hate, good and evil, are values subservient to the human moral predicament. As unique moral animals endowed with the capacity and the wherewithal to act according to reason, to steer our actions not just according to our wants and desires but also according to the directives of right and wrong, we human beings are equally empowered and burdened with the responsibility to justify our actions.

The Vietnam War and the early days of the environmental movement have in common the shaping of the American political psyche and the framing of political discourse. To the delight of political scientists, they share a similar structure. They are both characterizable as battles: one between the United States and the shadow soldiers of the Vietcong; the other between tree huggers and, take your pick: anti-environmentalists, big business, corporate capitalism and so on. This much is well established. They also share a tension in classic reaction formation. Where the Vietnam War was caricatured at the time as motivated by hatred of the enemy, the environmental movement has subsequently been caricatured as motivated by an inversion of this: love and kisses for the "enemy," hugs for nature. Indeed, the antiwar protesters were derided as communist dupes and peaceniks. What they were actually doing, however, was precisely what they should have been doing: arguing about the unjustifiability of the war.

The environmental movement has, for the most part, always striven to be a true velvet revolution. But as we have seen, those who understand the movement in this way, who think that environmentalism is about the love of nature or the good in nature, misunderstand the underpinnings of obligation as well as the nature of nature. It may historically be true that the Vietnam War was fueled by the raw anger of those who participated in it, or that tree huggers and hippies have been driven by an overwhelming love for the earth, but this descriptive fact does not alter what rests at the heart of the question of what we should do. This question, the ethical question, can be answered only by deference to the imperfect and flawed, but nevertheless flesh-and-blood justificatory apparatus that we already have in place in civil society.

The orientation toward success—underscored in the Vietnam conflict, amplified through the war on terrorism, highlighted by the repeated calls of environmentally concerned scientists and activists to turn the tide of catastrophe—is not merely ineffectual, as many critics allege; it is conceptually flawed, rooted in a misunderstanding of our obligations to one another. Humans are not, and have not been, at war with nature, and the corrective to such a mythologized adversarial relationship with nature is therefore not love. We needn't romance nature, nor lionize it, nor elevate it to the status of a goddess, to understand that we have some pretty stringent obligations to conserve or preserve it. These obligations emerge from the requirement that we routinely check to make sure that our actions are justifiable. A decent bet here is that a good number of our practices are not as easily justified as we often assume them to be. This is also true about our obligations to one another. We needn't love one another to do right by one another. We just need to make sure that we weigh the interests and needs of others by leaving our actions open to their scrutiny. They should do the same for us.

The orientation toward justification, as captured by the model of the Viridian Commonwealth, emphasizes fair and equal citizenship among like parties. It suggests that there is something special about humanity, a uniqueness that demands we do more than just act like forces of nature. It requires us to exercise our will, to live life like we mean it, to take responsibility for our actions. It permits that in some cases we'll need to utilize the resources at our feet, but that those cases, the appropriate cases, can be unearthed only by actively engaging in this important discussion about when it is and is not appropriate, or justified, or right, to use those resources.

To be just, to do right—in the end, to be good—we need to act with a reason that is good *and* justified. We must ask whether any person of any background in exactly the same circumstances and given more perfect knowledge would do what we're about to do. But we should go beyond even this. We must actively seek feedback on our reasons for acting. We've got to take steps to ensure that we don't live in a fantasyland where resources can be used recklessly to feed our wildest desires. We can do this by honestly and charitably inviting contributions to and criticism of our reasoning about what to do.

At the individual level, when each of us walks into a store, we must ask whether the purchase we intend to make is necessary and right. If our product is made by exploited or vulnerable people, if it will create a problem or destroy something of value, we should ask whether there is a better option, an equivalently good item that could reasonably serve as a substitute.

At the collective level, when voting or electing officials, we must ask whether a policy or a rule is necessary and right, whether the state or a suitably empowered authority is well positioned to redress problems created by individuals acting independently.

The burden of humanity is a heavy load indeed. We can, of course, live our lives without accepting this burden. It may seem like fun to take advantage of others and disrespect them. We can also quite happily live our lives by satisfying our most basic desires. In my opinion, down this brutish road lies not happiness but a state of perpetual grief, a life that is otherwise solitary, poor, and nasty. It is much better to be an unhappy but conscientious person, it has been said, than a blissful and brutish ignoramus.

This is the fundamental ecological conundrum. It would be one thing if we were destroying the earth for a good reason. That, perhaps, we could countenance. But we're not. We're destroying the earth for no reason.

It is not as if we're covering the landscape with hospitals and museums, creating civic spaces for each of us to grow and learn in, or constructing the next great society. We're not. We're producing widgets and durable goods we don't need and won't use, like clappers and beanie babies, that will slowly (if ever) decay in a landfill, smiling stupidly up at the sky for centuries after we've passed away.

I've made this point in ten chapters, slowly walking you through my reasoning. In chapter 1 I argued that catastrophe reasoning about climate change is the wrong way to think about our relationship to the environment. In chapter 2 I explored several ethical positions that seem to inform the environmental movement and focused on those orientations that specify the value of nature in order to motivate us to take proenvironment stances. In chapter 3 I introduced the notion that nature isn't all that it's cracked up to be, which I did by cataloguing the ways in which nature can be quite bad. In chapter 4 I introduced several comparison cases, reasoning that natural events should be kept separate from human actions.

This inaugurated the second section. In chapter 5 I dissected actions by proposing that they are distinct from events insofar as they are subject to the influence of reasons, and reasons of the will in particular. In chapter 6 I proposed that our determinations about the right course of action can't rely solely on a view about better and worse ends but must instead also involve some assessment of the reasons that steer us toward those ends. In chapter 7 I claimed that we already rely on these nonconsequentialist norms to steer us in many of our endeavors, like voting.

By the third section of the book, I was already heavily engrossed in the discussion of actions. I took a somewhat different tack in chapter 8 when I reframed the discussion of means and ends to emphasize two attitudinal orientations: the orientation toward success and the orientation toward justification. In chapter 9 I further fleshed out my reasoning for the appropriateness of the orientation toward justification by making an argument for the priority of the right over the good. And in chapter 10 I hoped to cement your agreement by having you envision three separate worlds, Wicked, Wild, and Earth.

• • •

Now I'd like you to try something:

1. When you have a bit of free time, go to a store. It needn't be a grocery store, but grocery stores are nice because they have such

tremendous inventories. Just find a place where you'll have to make a choice.

2. Stop in the aisle and pick a product, any product. Maybe something you wouldn't normally buy, something you're curious about, something that seems out of the ordinary, or something that is outrageously common.

3. Reflect on the contents of that product: Is it healthy? Is it good for you? Is it something that will satisfy a want? Will it satisfy a need? Will your life be better or worse if you consume that product? What is its price?

4. Think about what went into the production of that item. Where did it come from? How did it get to the store—by truck or by train? What's in it? How is the environment impacted by that product? Were any living beings harmed in the making of that product? Who made it? What were the conditions in which it was made? Do any of these considerations require further investigation?

5. If so, go home, get online, and research your product. Learn more about it. What political forces are in play to secure that the product makes it onto shelves? What is the total cost of production? What is the wholesale price? Is it cheaper at other stores? What accounts for the difference? These are all important questions.

6. If you're not exhausted yet, bring it up in conversation at dinner that evening. Do you know where your product comes from? Do you know what goes into its production? What must it be like to work in the factory or on the farm that produces that product? What does your family think?

This is the justificatory process that we should be engaging in every time we make a purchase. If we find it too exhausting, or too cumbersome, or too demanding, that's understandable, since there is only so much time in our day and our cognitive capacity is constrained by the amount of sleep we have gotten, the number and intensity of children pulling on our apron strings, and other things. But if we find this exhausting (and I certainly do), then we should insist on constraints that make this process easier for us. We can insist on these constraints by ensuring that we have collective-level policies in place that keep the justificatory work at the individual level to a manageable amount, and we should work to ensure that the institutions that set these policies are structured to promote collective-level justification.

• • •

When I first began this book, my son Jasper was four. He was a bundle of desires. He wanted anything that flashed, looked warm, felt fuzzy, beeped, or buzzed. He liked trucks. He *loved* books. He forged deep relationships with stuffed animals. He especially adored creatures with big eyes. And like many kids his age, he had a menagerie of hard-worn plastic dinosaurs that he clobbered, every now and then, with a miniature space rock—or a ball, or a block, or whatever too-heavy object happened to be lying nearby—as if to reenact a 64-million-year-old tale that could not possibly have been told to him in any detail by even the most elementary playground paleontologist.

If we were in a store, he might see some lovelorn creature on the shelf and make sure to let me know that he wanted it. He'd press for objects that he thought he needed, even though the kindness of friends and parents had gifted him with a mountain of recycled, once-loved animals that towered over him as he slept. Though he was only four, his mother and I asked him to justify his requests. We asked him to give us reasons. If he answered that he "just wants it," we'd deny that this was a satisfactory reason. If he gave us some other reason, even something very small, we would listen and make a determination with him about the appropriateness of the purchase. Always, we wanted to know if he thought it right for him to get that item—right then, right there. Often, we didn't buy anything.

It may seem somewhat unfair to request a four-year-old to articulate reasons, but I've seen many parents do similar things with their own children, each in their own ways. They do not allow their kids to eat the syrupy treats that punctuate the grocery store checkout lines. They do not give in when their kids throw a tantrum. Like us, they require that their kids act within reason. Sometimes they take the time to explain to their children why they cannot have what they want, but just as often they don't bother, and instead make snap determinations of whether a request is justified. It's a delicate balancing act. Each parent has her own style, and each, I am sure, frets about being too easy or too harsh.

Though he is older now, Jasper is a very happy kid. He is well-adjusted and friendly. He treats inanimate objects with care and stares with astonishment at all living creatures. Who knows? Perhaps Jasper will grow to crave the immediate gratification that his oppressive parents have denied him. But I suspect he won't. I suspect he'll mature into a thoughtful boy, and eventually into a reflective and contributing member of society. At least, it is my deepest wish that he will.

Just as we do with our children, so too must we do with ourselves. We must institute policies of rational restraint, asking whether our consumption of the latest buzzing, flashing, big-eyed-animal-on-the-shelf is really what we need—right then, right there; asking whether our desires can withstand the scrutiny of others, whether we can justify our actions to our neighbors and to our children's children. Far from the environmental fantasy of a starry-eyed idealist, this is a steep moral requirement. It demands of us a kind of moral maturity, an ethical sophistication not to give in to our animal desires, and instead to grow up. Being green does not require us to love nature, or cherish it, or wax romantic over some far-off bucolic ecotopia. Rather, it means that we must grow out of our childish natures, stop salivating at the checkout counter, and do the right—the justified—thing.

Notes

INTRODUCTION

1. As with any social movement, there are many diagnoses for what ails environmentalism. Michael Shellenberger and Ted Nordhaus have argued that environmentalism has flagged because it is fundamentally a movement situated in fear and despair; see their famous *Death of Environmentalism*, 2004, originally self-published but now featured at: http://www.thebreakthrough.org/images/Death_of_Environmentalism.pdf.

2. Rush Limbaugh, "Environmentalist Wacko: Climate Change Skeptics Are Sick," April 2, 2012, http://www.rushlimbaugh.com/daily/2012/04/02/environmentalist_wacko_climate_change_skeptics_are_sick.

3. Kate Sheppard, "Ted Cruz: 'Global Warming Alarmists Are the Equivalent of the Flat-Earthers,'" *Huffington Post*, March 26, 2015, http://www.huffingtonpost.com/2015/03/25/ted-cruz-global-warming_n_6940188.html.

4. Quoting Krauthammer: "One reason that the drilling is happening in the Gulf that deep is his [Obama's] allies on the left aren't going to allow it on the intercontinental shelf, where it's more safe, and in the arctic, where we know how to do it and where if you were to have a spill it would injure seals and caribou but not humans, as is happening on the Gulf coast of the United States." Fox News, *Special Report*, May 14, 2010.

5. Some very astute readers may bristle at my humanism here, charging me with promoting human exceptionalism. I think that's not exactly correct, though I do believe that environmentalism is a humanism. Inasmuch as my position is that the right has priority over the good, which is to say that the good is justified by the right, all claims of betterness or worseness are always already filtered through the lens of the right. So, all things considered, a justified action culminating in some outcome is, ethically speaking, better than a natural event culminating in the same outcome. This is true also of justified actions versus unjustified (mere) actions. My statements therefore are not intended to suggest that humans are better *in kind* than nature; but only that since we humans can justify our actions, we can be better than nature.

6. Rachel Weiner, "Obama: 'We Can't Solve Global Warming Because I F—ing Changed Light Bulbs in My House," *Huffington Post*, May 25, 2011, http://www.huffingtonpost.com/2008/11/05/obama-we-cant-solve-globa_n_141407.html.

CHAPTER 1: RETURN TO THE PALEOCENE

1. Arthur Schopenhauer, "On the Sufferings of the World," in *Studies in Pessimism* (New York: Cosimo, 2007), 5–6.

2. Christopher McGowan, *The Dragon Seekers: How An Extraordinary Circle of Fossilists Discovered the Dinosaurs and Paved the Way for Darwin* (New York: Perseus Books, 2001); Deborah Cadbury, *Terrible Lizard: The First Dinosaur Hunters and the Birth of a New Science* (New York: Henry Holt, 2000).

3. Martin J. S. Rudwick, *The Meaning of Fossils: Episodes in the History of Paleontology* (Chicago: University of Chicago Press, 1976); Dennis R. Dean, *Gideon Mantell and the Discovery of Dinosaurs* (Cambridge, UK: Cambridge University Press, 1999); J. C. T. Fairbank, "William Adams and the Spine of GA Mantell," *Annals of the Royal College of Surgeons of England* 86 (2004): 349–352.

4. Scott Wing et al., "Transient Floral Change and Rapid Warming at the Paleocene-Eocene Boundary," *Science* 310, no. 5750 (November 11, 2005): 993–996.

5. While this is an important hypothesis, recent work in this area suggests that carbon dioxide may not play as strong a forcing role as some scientists have claimed; this lends further credence to the ethical position I advance in this book, that getting the science right is only one piece in a giant justificatory puzzle. See, e.g., Richard Zeebe et al., "Carbon Dioxide Forcing Alone Insufficient to Explain Paleocene-Eocene Thermal Maximum Warming," *Nature Geoscience* 2, no. 8 (2009): 576–580, http://www.nature.com/ngeo/journal/v2/n8/abs/ngeo578.html.

6. Peter Ward, *Under a Green Sky: Global Warming, the Mass Extinctions of the Past, and What They Can Tell Us about Our Future* (New York: Harper Perennial, 2008).

7. Immanuel Kant, *Critique of Pure Reason*, trans. Norman Kemp Smith (1781; repr., New York: St. Martin's Press, 1965), A51–B75.

8. Intergovernmental Panel on Climate Change (IPCC), "Summary for Policymakers," in *Climate Change 2014: Impacts, Adaptation, and Vulnerability* (Cambridge, UK: Cambridge University Press, 2014), 1–32, http://www.ipcc.ch/report/ar5/wg2.

9. Max Boykoff and Jules Boykoff, "Climate Change and Journalistic Norms: A Case Study of US Mass-Media Coverage," *Geoforum* 38 (2007): 1190–1204; Maxwell T. Boykoff, "We Speak for the Trees: Media Reporting on the Environment," *Annual Review of Environment and Resources* 34 (November 2009): 431–457.

10. Andrew Revkin, "Climate Change as News: Challenges in Communicating Environmental Science," in *Climate Change: What It Means for Us, Our Children, and Our Grandchildren*, ed. J. F. C. DiMento (Cambridge, MA: MIT Press, 2007), 139–159. Credit for pulling the quote goes to Max Boykoff.

11. It can, of course, be argued that mathematical and logical arguments are normatively "pure" in the sense that they don't carry a prescriptive valence. This is why I say that scientific arguments as a class only *frequently* carry this valence.

12. There's a fair amount of literature on this; see, e.g., Elijah Millgram, ed., *Varieties of Practical Reasoning* (Cambridge, MA: MIT Press. 2001).

13. IPCC, "Summary for Policymakers," 8–10.

14. Ibid., 20.

15. Fox News, *Your World with Neil Cavuto*, December 14, 2015, Media Matters, http://www.mediamatters.org/video/2015/12/14/ben-stein-attacks-global-climate-agreement-savi/207486.

16. George Will, "Pope Francis's Fact-Free Flamboyance," *Washington Post*, September 18, 2015.

17. George Will, "Let Cooler Heads Prevail," *Washington Post*, April 2, 2006.

18. The dominance of uniformitarianism is relatively difficult to establish. How, exactly, can one claim that a certain view is "dominant"? Nevertheless, this is the claim made by Martin Rudwick, *The Meaning of Fossils: Episodes in the History of Paleontology* (Chicago: University of Chicago Press, 1985). Rudwick doesn't attribute the resurgence in catastrophism to the publication of the Alverez asteroid article per se, but the article was unquestionably instrumental in the resurgence. This was argued by Trevor Palmer, "The Rise and Fall of Catastrophism," lecture presented at the University of Nottingham, April 25, 1996.

19. Garrett Hardin, "The Tragedy of the Commons," *Science* 162 (1958): 1243–1247.

20. Matthew Nisbet, a professor of communication at Northeastern University, has been making this point for some time now. He notes that "urgent calls to escalate the war against climate skeptics may lead scientists and their organizations into a dangerous trap, fueling further political disagreement while risking public trust in science." But he takes

issue with this approach on political grounds. According to his reasoning, it's unnecessary for climate scientists and environmentalists to revert to a "bunker mentality." The climate scientists are winning. Poll after poll demonstrates that about 75 percent of the American public continues to trust climate science, despite the attacks from skeptics. Matthew Nisbet, "Chill Out: Climate Scientists Are Getting a Little Too Angry for Their Own Good," *Slate*, March 18, 2010, http://www.slate.com/articles/health_and_science/green_room/2010/03/chill_out.html.

21. Naomi Oreskes and Erik M. Conway, *Merchants of Doubt: How a Handful of Scientists Obscured the Truth on Issues from Tobacco Smoke to Global Warming* (New York: Bloomsbury Press, 2010).

22. George Lakoff, "Pope Francis Gets the Moral Framing Right: Global Warming Is Where the Practical and the Moral Meet," *Huffington Post*, June 25, 2015, http://www.huffingtonpost.com/george-lakoff/pope-francis-gets-the-mor_b_7665694.html.

23. Alan Harris, "The Odds of an Asteroid Strike," *Nova Next*, PBS, March 27, 2013. http://www.pbs.org/wgbh/nova/next/space/risk-of-an-asteroid-strike.

CHAPTER 2: THE PRECIOUS VASE

1. To be fair here, and also to protect the innocent, I took some poetic license with a few of the names.

2. Some may take issue with the claim that somehow Americans are now more concerned about the environment than they were in the past, and perhaps for good reason. They may point to various accounts that demonstrate a long-standing history of interest in nature. See, e.g., Roderick Nash, *Wilderness and the American Mind* (New Haven, CT: Yale University Press, 1982); Thomas Dunlap, *Saving America's Wildlife: Ecology and the American Mind, 18550–1990* (Princeton, NJ: Princeton University Press, 1991); John Gatta, *Making Nature Sacred: Literature, Religion, and Environment in America from the Puritans to the Present* (Oxford, UK: Oxford University Press, 2004); John F. Sears, *Sacred Places: American Tourist Attractions in the Nineteenth Century* (Amherst, MA: University of Massachusetts Press, 1999); and Ben Minteer, *Landscape of Reform: Civic Pragmatism and Environmental Thought in America* (Cambridge, MA: MIT Press, 2009). I only mean to point out anecdotally that as a point of political concern, the environmental movement is a relatively young movement. It is true that authors have been concerned about nature for decades and also that the strength of the average commitment to environmental concerns waxes and wanes with the years.

3. See, e.g., the websites of the World Wildlife Fund (http://www.worldwildlife.org), the Wilderness Society (http://wilderness.org), the Sierra Club (http://www.sierraclub.org), and the Nature Conservancy (http://www.nature.org/ourinitiatives/urgentissues/climatechange/index.htm).

4. See, e.g., the websites of Friends of the Earth (http://www.foe.org/projects/climate-and-energy) and Greenpeace (http://www.greenpeace.org/usa/en).

5. See, e.g., the websites of 350 (http://www.350.org) and the Bellona Foundation (http://www.bellona.org).

6. Roger Pielke Jr., *The Climate Fix: What Scientists and Politicians Won't Tell You about Global Warming* (New York: Basic Books, 2010).

7. Juliet Eilperin, "Inhofe's Al Gore Igloo," *Washington Post*, February 9, 2010.

8. Gavin Schmidt and Stefan Rahmstorf. "Uncertainty, Noise and the Art of Model-Data Comparison," Real Climate, January 11, 2008, http://www.realclimate.org/index.php/archives/2008/01/uncertainty-noise-and-the-art-of-model-data-comparison.

9. Aristotle, *Nicomachean Ethics*, 1098a. Translations vary widely.

10. See, e.g., "Tips on Communicating with News Media," Yale Climate Connections, December 15, 2014, http://www.yaleclimateconnections.org/2014/12/tips-on-communicating-with-news-media.

11. This point has been made repeatedly by commentators on the environmental community. More recently, however, as the prospect of successful climate legislation has waned, at least in the United States, the environmental community has turned its attention to related, but nevertheless more localized, battles, such as the BP oil spill in the Gulf of Mexico and the Japanese disaster at the Fukushima Daiichi nuclear power plant.

12. Often, but not always, the extrinsic value position is associated with anthropocentrism, whereas the intrinsic value position is associated with nonanthropocentrism, but I think that using these labels can be more confusing than helpful.

13. Christopher D. Stone, *Should Trees Have Standing?: Law, Morality, and the Environment*, 3rd ed. (New York: Oxford University Press, 2010).

14. I am using the term *Darwinian* here colloquially. Charles Darwin did acknowledge the importance of social cooperation to survival.

15. Aldo Leopold, *A Sand County Almanac: With Other Essays on Conservation from Round River* (New York: Oxford University Press, 1966).

16. See Murray Bookchin, "Social Ecology versus Deep Ecology: A Challenge for the Ecology Movement," June 25, 1987, at http://dwardmac.pitzer.edu/Anarchist_Archives/bookchin/socecovdeepeco.html.

17. Graciela Chichilnisky and Geoffrey Heal, "Economic Returns from the Biosphere," *Nature* 391 (1998): 629–630.

18. Robert Costanza et al., "The Value of the World's Ecosystem Services and Natural Capital," *Nature* 387 (1997), 253–260.

19. Gretchen Daily and Katherine Ellison, *The New Economy of Nature: The Quest to Make Conservation Profitable* (Washington, DC: Island Press, 2003).

20. To my knowledge, the best arguments against this position can be found in Mark Sagoff, *The Economy of the Earth: Philosophy, Law, and the Environment* (Cambridge, UK: Cambridge University Press, 2007); and Mark Sagoff, *Price, Principle, and Environment* (Cambridge, UK: Cambridge University Press, 2004).

21. Nicholas Stern, *The Economics of Climate Change: The Stern Review* (Cambridge, UK: Cambridge University Press, 2007).

22. Thanks to economist Eban Goodstein of Lewis and Clark College for this point.

23. See Richard S. J. Tol, "The Economic Effects of Climate Change, *Journal of Economic Perspectives* 23, no. 2 (2009): 29–51; and William Nordhaus, The *Stern Review* on the Economics of Climate Change," May 3, 2007, http://www.econ.yale.edu/~nordhaus/homepage/stern_050307.pdf.

CHAPTER 3: RUSTLING IN THE BUSHES

1. "Preventing Dog Bites," Centers for Disease Control, May 18, 2015, http://www.cdc.gov/features/dog-bite-prevention.

2. "Animal Bites," World Health Organization, February 2013, http://www.who.int/mediacentre/factsheets/fs373/en.

3. R. L. Langley and W. E. Morrow, "Deaths Resulting from Animal Attacks in the United States," *Wilderness Environmental Medicine* 8, no. 1 (1997): 8–16.

4. Paul Beier, "Cougar Attacks on Humans in the United States and Canada," *Wildlife Society Bulletin* 19, no. 4 (1991): 403–412.

5. See http://articles.orlandosentinel.com/2006-05-12/news/GATOR12_1_alligator-attack-perper-broward.

6. Malcolm Brown, "After 32 Years of Speculation, It's Finally Official: A Dingo Took Azaria." *Sydney Morning Herald*, June 12, 2012.

7. Edward Abbey, *A Voice Crying in the Wilderness (Vox Clamantis in Deserto): Notes from a Secret Journal* (New York: St. Martin's Press, 1991).

8. Benjamin Radford, *Tracking the Chupacabra: The Vampire Beast in Fact, Fiction, and Folklore* (Albuquerque: University of New Mexico Press, 2011), 172.

9. Haden Blackman, *The Field Guide to North American Monsters: Everything You Need to Know about Encountering Over 100 Terrifying Creatures in the Wild* (New York: Three Rivers Press, 1998).

10. Aristotle, *Historia Animālium*, trans. D'Arcy Wentworth Thompson (Adelaide, Australia: University of Adelaide, 2005), e-book.

11. Charles T. Wolfe, ed., *Monsters and Philosophy* (London: Kings College, 2005).

12. Francis Bacon, "Preparative toward Natural and Experimental History" (Cambridge, UK: Cambridge University Press, 2011).

13. By which Hobbes means, equal capacity to outwit and overpower one another.

14. Jean-Jacques Rousseau, *Émile, or On Education*, trans. Allan Bloom (New York: Basic books, 1979), 37.

15. "Facts about Herbicides," US Department of Veterans Affairs, http://www.publichealth.va.gov/exposures/agentorange/basics.asp.

16. Clyde Haberman, "Agent Orange's Long Legacy, for Vietnam and Veterans," *New York Times*, May 11, 2014.

17. "Veterans' Diseases Associated with Agent Orange," US Department of Veterans Affairs, http://www.publichealth.va.gov/exposures/agentorange/diseases.asp#veterans.

18. E. O. Wilson, "Afterword," in *Silent Spring*, by Rachel Carson (New York: Houghton-Mifflin, 2002).

19. Kirsten Weir, "Rachel Carson's Birthday Bashing," *Salon*, June 29, 2007, http://www.salon.com/news/feature/2007/06/29/rachel_carson.

20. Peter Matthiessen, "Environmentalist Rachel Carson," *Time*, March 29, 1999, http://www.time.com/time/magazine/article/0,9171,990622-3,00.html.

21. Editorial, "Forty Years of Perverse 'Responsibility,'" *Washington Times*, April 28, 2007, http://www.washingtontimes.com/news/2007/apr/28/20070428-100957-5274r.

22. Monsanto is mostly involved in the development of genetically modified seed. Other entrepreneurial ventures, such as AquaBounty Technologies, have sought to modify animals for food and organ production. See "Genetically Engineered Animals," US Food and

Drug Administration, http://www.fda.gov/animalveterinary/developmentapprovalprocess/geneticengineering/geneticallyengineeredanimals/default.htm.

23. Jennifer Viegas, "Chupacabra Mystery Solved," Discovery, October 22, 2010. http://news.discovery.com/animals/pets/chupacabra-mystery-solved.htm.

24. The philosopher Martin Heidegger proposes that this is the core danger of modern technology. It frames the world in such a way that it all appears to be a sort of standing reserve. Martin Heidegger, *The Question Concerning Technology* (New York: Harper and Row, 1977).

CHAPTER 4: THE WILD AND THE WICKED

1. Some scholars have recently challenged this view quite aggressively. Tsuyoshi Hasegawa, a professor of history at the University of California, Santa Barbara, has written that it was the Soviet entry of the war in the Pacific that eventually forced Japan's hand, not the destruction of Hiroshima and Nagasaki. He also quotes the Pulitzer Prize–winning author Richard Rhodes, who wrote the 1986 book *The Making of the Atomic Bomb*, as backing down on the view that the bombings ended the war. Tsuyoshi Hasegawa, "Were the Atomic Bombings of Hiroshima and Nagasaki Justified?," in *Bombing Civilians: A Twentieth-Century History*, ed. Yuki Tanaka and Marilyn Young (New York: New Press, 2009), 97–134. See also Gareth Cook, "Why Did Japan Surrender?" Boston.com, August 7, 2011, http://www.boston.com/bostonglobe/ideas/articles/2011/08/07/why_did_japan_surrender/?page=full.

2. Interview with J. Robert Oppenheimer about the Trinity explosion, first broadcast as part of the television documentary *The Decision to Drop the Bomb* (1965), produced by Fred Freed, NBC White Paper.

CHAPTER 5: CONTROL FREAK

1. Although it is documented in numerous places as the Wapiti Shelter, and presumably the shelter is named after the wapiti, which is an elk, some sources do refer to it as the Wapitu Shelter.

2. Chris Hansen, "Escape from Brushy Mountain," *Dateline NBC*, MSNBC, February 15, 2009, http://www.msnbc.msn.com/id/29187510.

3. Wil Haygood, "Blood on the Mountain," *Washington Post*, July 8, 2008, http://www.washingtonpost.com/wp-dyn/content/article/2008/07/07/AR2008070702332_pf.html.

4. Jeremy Bentham, *The Principles of Moral Legislation* (Amherst, NY: Prometheus Books, 1988), originally published in 1780.

5. Haygood, "Blood."

6. Hansen, "Escape."

7. Haygood, "Blood."

8. Careful readers will perhaps object here that I am conflating at least two views: on will and freedom of the will. There is an extremely complicated and dense body of literature on this topic. For our purposes, however, I don't think we need to be too concerned with the specifics of any particular theory. What's important, it seems to me, is to get a general sense of what the will might be.

9. See Harry Frankfurt, "Freedom of the Will and the Concept of the Person," *Journal of Philosophy* 68, no. 1 (1971): 5–20.

CHAPTER 6: DR. FEELGOOD AND MR. FIX-IT GO TO THE PICTURE SHOW

1. Portions of this section appeared as an article for the online journal *Science Progress*. In that earlier version I argued primarily along rights lines. Here I will argue along responsibility and respect lines. Benjamin Hale, "You Say 'Solution,' I Say 'Pollution': Ocean Fertilization Is a Fishy Solution to a Whale of a Problem," *Science Progress*, August 18, 2009, http://www.scienceprogress.org/2009/08/ocean-fertilization-ethics.

2. See, e.g., H. E. Willoughby, D. P. Jorgensen, R. A. Black, and S. L. Rosenthal, "Project STORMFURY: A Scientific Chronicle, 1962–1983," *Bulletin of the American Meteorological Society*, 66, no. 5 (1985), 505–514.

3. In the case of parts per million or two-degree targets, we're not so much aiming for the target as seeking to constrain our behavior so that we limit the damage from our behavior. But it's the target or goal setting that interests me, and in these cases the target is an attempt not to create a better world but to avoid a worse world.

4. Kelsey Campbell-Dollaghan, "Celebration, Florida: The Utopian Town That America Just Couldn't Trust," Gizmodo, April 20, 2014, http://gizmodo.com/celebration-florida-the-utopian-town-that-america-jus-1564479405; see also Andrew Ross, *The Celebration Chronicles: Life, Liberty, and the Pursuit of Property Value in Disney's New Town* (New York: Ballantine Books, 2000).

5. "Environment: A Dome for Winooski?", *Time*, December 10, 1979.

6. "Millennium Dome," Politics, http://www.politics.co.uk/reference/millennium-dome.

7. For more on this thesis, see James C. Scott, *Seeing Like a State: How Certain Schemes to Improve the Human Condition Have Failed* (New Haven, CT: Yale University Press, 1999).

8. Vladimir Lenin, *Collected Works, Vol. 18: April 1912–March 1913* (Moscow: Progress Publishers, 1963), 356.

9. See, e.g., Tim Ball, "Fighting the Wrong Battle: Public Persuaded about CO_2 as Pollutant—Not as Cause of Warming," Watts Up with That, February 5, 2014, http://wattsupwiththat.com/2014/02/05/fighting-the-wrong-battle-public-persuaded-about-co2-as-pollutant-not-as-cause-of-warming.

10. Ross, *Celebration Chronicles*.

11. Michael Pollan, "Sticks and Stones: Mickey for Mayor? Disney's New Town May Be So Perfect It's a Nightmare," *House & Garden*, October 1996.

12. Jennifer Brown, "Children's Illness Stirs Debate on Raw Milk," *Denver Post*, September 6, 2010, http://www.denverpost.com/search/ci_16001821; Vanessa Miller, "Longmont's Billy Goat Dairy Allowed to Reopen after Sickening 30," *Boulder Daily Camera*, July 15, 2010, http://www.dailycamera.com/ci_15522785?IADID.

CHAPTER 7: THE VOTER'S CONUNDRUM

1. Rachel E. Dwyer, "Expanding Homes and Increasing Inequalities: US Housing Development and the Residential Segregation of the Affluent," *Social Problems* 54 (2007): 23–46.

2. Thom File and Sarah Crissey, "Voting and Registration in the Election of November 2008," US Bureau of the Census, July 2012, http://www.census.gov/prod/2010pubs/p20-562.pdf.

3. In philosophy, this is known as the *causal impotence* problem. I and many other environmental ethicists have written on this topic. See, e.g., Benjamin Hale, "Nonrenewable Resources and the Inevitability of Outcomes," *Monist* 94, no. 1 (July 2011): 369–390; and Walter Sinnott-Armstrong, ed., "It's Not My Fault: Global Warming and Individual Moral Obligations," in *Perspectives on Climate Change: Science, Economics, Politics, Ethics*, 285–307 (Amsterdam: Elsevier, 2005).

4. "The Story of Non-Economist Elinor Ostrom," *Swedish Wire*, December 9, 2009, http://www.swedishwire.com/business/1985-the-story-of-non-economist-elinor-ostrom.

5. For a good discussion of this, see Robert Pippin, "Rüdiger Bittner: Doing Things for Reasons," Notre Dame Philosophical Reviews, http://ndpr.nd.edu/review.cfm?id=1130.

6. Richard Feynman, "Fun to Imagine," BBC, July 15, 1983, http://www.bbc.co.uk/archive/feynman/10701.shtml.

CHAPTER 8: THE AXES OF EVILDOERS

1. Robert Boynton, "Powder Burn: Eco-Terrorism at Vail," *Outside*, January 1999.

2. "Bush Vows to Rid the World of 'Evildoers,'" CNN, September 16, 2001, http://edition.cnn.com/2001/US/09/16/gen.bush.terrorism.

3. Kelly Hearn, "Stepping Up the Attack on Green Activists," Alternet, September 29, 2005, http://www.alternet.org/environment/26077; Will Potter, "Lobbying Documents Show How Corporations Snuck 'Eco-Terrorism' Law through Congress," Green Is the New Red, December 19, 2008, http://www.greenisthenewred.com/blog/lobbying-documents-show-how-corporations-snuck-eco-terrorism-law-through-congress-part-2-of-3/820.

4. Sara Burnett, "Reward in Vail's Two Elk Lodge Arson," *Rocky Mountain News*, November 20, 2008.

5. Tracy L. Dietz, "An Examination of Violence and Gender Role Portrayals in Video Games: Implications for Gender Socialization and Aggressive Behavior," *Sex Roles* 38, nos. 5–6 (March 1998): 425–442.

6. "The Culture of Gaming," *On the Media*, NPR, December 31, 2010, http://www.wnyc.org/story/133034-the-culture-of-gaming.

7. "The Future of Gaming," *On the Media*, NPR, December 31, 2010, http://www.wnyc.org/story/143600-future-gaming/#transcript.

8. "Resolution on Violent Video Games," American Psychological Association, 2015, http://www.apa.org/news/press/releases/2015/08/violent-video-games.pdf; Craig A. Anderson, "Violent Video Games: Myths, Facts, and Unanswered Questions," American Psychological Association, October 2003, http://www.apa.org/science/about/psa/2003/10/anderson.aspx.

9. Craig A. Anderson, Douglas A. Gentile, and Katherine E Buckley, *Violent Video Game Effects on Children and Adolescents: Theory, Research, and Public Policy* (Oxford, UK: Oxford University Press, 2007).

11. Steve Lipsher, "Guilty Pleas Unveil the Tale of Eco-Arson on Vail Summit," *Denver Post*, December 15, 2006.

12. Kevin Tubbs, "A Statement from Kevin," Support Kevin Tubbs, http://web.archive.org/web/20120420111322/http://www.supportkevintubbs.com/.

13. Mike Godwin, "Meme, Countermeme." *Wired*, October 1, 1994, http://www.wired.com/1994/10/godwin-if-2.

14. Ben Fishel, "Beck on *An Inconvenient Truth*: 'It's Like Hitler,'" Media Matters, June 8, 2006, http://mediamatters.org/mmtv/200606080005; Jonathan H. Adler, "Lomborg = Hitler?", *National Review*, May 11, 2004,

https://web.archive.org/web/20140316174001/http://www.nationalreview.com/corner/79444/lomborg-hitler/jonathan-h-adler; Judd Legum, "Sen. Inhofe Compares People Who Believe in Global Warming to 'the Third Reich,'" *Climate Progress*, July 24, 2006, http://thinkprogress.org/2006/07/24/inhofe-third-reich; Andrew Revkin, "Climate, Coal, and Crematoria," *New York Times*, November 26, 2007, http://dotearth.blogs.nytimes.com/2007/11/26/holocausts; David Roberts, "Is the Analogy between Climate Change and Hitler's Atrocities Appropriate?" *Grist*, November 7, 2007. http://www.grist.org/article/global-warming-and-the-holocaust; John Vidal, "Viscount Monckton Calls Young Climate Activists 'Hitler Youth,'" *Guardian,* December 11, 2009, http://www.guardian.co.uk/environment/blog/2009/dec/11/monckton-calls-activists-hitler-youth.

15. Julian Sanchez, "Frum, Cocktail Parties, and the Threat of Doubt," March 26, 2010, http://www.juliansanchez.com/2010/03/26/frum-cocktail-parties-and-the-threat-of-doubt.

16. See, e.g., the work of Dan Kahan at the Cultural Cognition Project, http://www.culturalcognition.net.

CHAPTER 9: THE GREEN'S GAMBIT

1. Brian Montopoli, "SC Lt. Gov. Andre Bauer Compares Helping Poor to Feeding Stray Animals," CBS, January 25, 2010, http://www.cbsnews.com/news/sc-lt-gov-andre-bauer-compares-helping-poor-to-feeding-stray-animals.

2. Garrett Hardin, "Lifeboat Ethics: The Case against Helping the Poor," *Psychology Today*, September 1974.

3. Shawn McCarthy, "Ted Turner Urges Global One-Child Policy to Save Planet," *Globe and Mail*, December 5, 2010, http://www.theglobeandmail.com/news/national/ted-turner-urges-global-one-child-policy-to-save-planet/article1825977.

4. Charles H. Eccleston, *NEPA and Environmental Planning: Tools, Techniques, and Approaches for Practitioners* (Boca Raton, FL: CRC Press, 2008), 120.

5. Linda Greenhouse, "Attack on Clean Air Act Falters in High Court Arguments," *New York Times,* November 8, 2000, http://query.nytimes.com/gst/fullpage.html?res=9F0DE4D61139F93BA35752C1A9669C8B63.

6. "The Rhino Hunter," *Radiolab*, NPR, September 7, 2015, http://www.radiolab.org/story/rhino-hunter.

7. Dana Milbank, "Civil Rights' New 'Owner': Glenn Beck," *Washington Post*, August 29, 2010, http://www.washingtonpost.com/wp-dyn/content/article/2010/08/27/AR2010082702359.html?hpid=opinionsbox1.

CHAPTER 10: THE GOOD GREEN LIFE

1. Joseph Fourier, "Remarques générales sur les températures du globe terrestre et des espaces planétaires," *Annales de Chimie et de Physique* 27 (1824): 136–167, https://geosci.uchicago.edu/~rtp1/papers/Fourier1827Trans.pdf.

2. David Darling, *The Universal Book of Mathematics: From Abracadabra to Zeno's Paradoxes* (Hoboken, NJ: John Wiley & Sons, 2004), 123–124.

3. R. M. Hare, "What Is Wrong with Slavery?" *Philosophy and Public Affairs* 8 (1979): 103–121.

4. "Attempts to Justify Slavery," Ethics Guide, BBC, http://www.bbc.co.uk/ethics/slavery/ethics/justifications.shtml.

5. Tim Lambert, "The Great DDT Hoax," *Deltoid*, Science Blogs, February 17, 2005, http://scienceblogs.com/deltoid/2005/02/17/ddt3/; Paul Driessen, *Eco-Imperialism: Green Power, Black Death* (New York: Merril Press, 2003).

6. "10 Facts on Malaria," World Health Organization, November 2015, http://www.who.int/features/factfiles/malaria/en/index.html.

7. John Sullivan, "$25,000 for a Hot Chocolate?" *New York Times*, November 7, 2007, http://cityroom.blogs.nytimes.com/2007/11/07/25000-for-a-hot-chocolate/index.html?hp.

• • •

Astute readers will recognize several passages in the text as closely resembling famous lines from other work, including, among other things, opening stanzas of Vladimir Nabokov's *Lolita* (New York: Random House, 1997) and D. H. Lawrence's *Lady Chatterley's Lover* (New York: Bantam Books, 1983). These similarities were intended primarily as an homage to great authors, not as any attempt to replicate their work without authorization. In the interest of integrity, here are the citations of relevance. Page 17: "Without further ado, ..." "... exhibit number one, ..." "... Ladies and gentlemen of the jury, ..." "simple, misinterpreted, ..." and "Look at this tangle of thorns," all stem from Nabokov, *Lolita*. Page 37: "Preliminary Expectorations," from Søren Kierkegaard, *Fear and Trembling / Repetition: Kierkegaard's Writings*, vol. 6 (Princeton: Princeton University Press, 1983). Page 43: "... terrible, horrible, no good, very bad, ..." from Judith Viorst, *Alexander and the Terrible, Horrible, No Good, Very Bad Day* (New York: Atheneum, 1987). Page 93: "no such thing as a free lunch," from Robert Heinlein, *The Moon Is a*

Harsh Mistress (New York: Orb Books, 1966). Pages 95–96: "Ours is essentially a tragic age, [but for years we have refused] to take it tragically, ..." "... new little habitats, ..." "... no smooth road into the future, ..." all from Lawrence, *Lady Chatterley's Lover*. Page 133: "something eminently human beaconing from his eye," from Robert Louis Stevenson, *The Strange Case of Dr. Jekyll and Mr. Hyde* (New York: Dover, 1991).

INDEX